DIALOGUES IN DATA POWER

Shifting Response-abilities in a Datafied World

Edited by
Juliane Jarke and Jo Bates

BRISTOL
UNIVERSITY
PRESS

First published in Great Britain in 2024 by

Bristol University Press
University of Bristol
1–9 Old Park Hill
Bristol
BS2 8BB
UK
t: +44 (0)117 374 6645
e: bup-info@bristol.ac.uk

Details of international sales and distribution partners are available at bristoluniversitypress.co.uk

British Library Cataloguing in Publication Data
A catalogue record for this book is available from the British Library

ISBN 978–1–5292–3830–3 paperback
ISBN 978–1–5292–3831–0 ePub
ISBN 978–1–5292–3832–7 OA PDF

Cover design: Nicky Borowiec
Front cover image: Getty Images/Klaus Vedfelt
Bristol University Press uses environmentally responsible print partners.
Printed and bound in Great Britain by CPI Group (UK) Ltd, Croydon, CR0 4YY

FSC
www.fsc.org
MIX
Paper | Supporting
responsible forestry
FSC® C013604

Contents

List of Figures

Notes on Contributors

Doris Allhutter is a senior scientist at the Institute of Technology Assessment, Austrian Academy of Sciences, where she leads the Austrian team of an international comparative study on 'Automating Welfare'. Her research focuses on the automation and datafication of Europe's systems for social protection, and on the implicit normativity of computing practices in machine learning under the lens of how these practices are entrenched in power relations. She is a Member of the Advisory Board on the Ethics of Artificial Intelligence of the Austrian UNESCO Commission.

Gaia Amadori is a PhD candidate in the Department of Communication, Università Cattolica del Sacro Cuore. Based on the DataChildFutures project, her dissertation studies the impact of data relations on the household's power structure through a mixed-methods research design. Her research interests concern media sociology, game studies, digital methods, and social network analysis.

Katrin Amelang is a postdoctoral researcher at the RUSTlab Bochum, Germany. She works at the intersection of cultural anthropology and (feminist) science and technology studies (STS), especially in the fields of (bio-)medicine and human-technology relations. Her current research explores the quantification/datafication of health and the body as well as everyday encounters with data, algorithms, and software.

Jo Bates is Professor of Data and Society at the University of Sheffield. Her research examines the social and cultural dynamics shaping data flows and practices, in sectors such as pharmaceuticals, higher education, climate science, and the arts. She works collaboratively with other researchers in critical data studies and related fields, as well as computer and information scientists who are engaged in the algorithmic processing of data.

Yaya Baumann is a PhD student in the Université de Montréal's Geography Department and an activist studying the potential of (digital) data and information technologies to foster community relations, to promote housing rights, and to support direct actions. Outside academia, Yaya likes to spend time in protests and programming. They also have a cat named Achille that they deeply love.

Caitlin Bentley is a lecturer in AI Education at King's College London. She is a co-director of the Safe and Trusted AI Centre for Doctoral Training and a member of the Responsible AI UK Delivery Team.

Hadley Beresford, University of Sheffield, is a postdoctoral researcher whose work is situated between information studies and sociological studies, and has been investigating how civil service departments can mitigate algorithmic bias in their services. Their doctoral research involved interviewing data scientists from the Department of Work and Pensions (DWP) in the UK, as well as practitioners on the Aurora AI project by the Finnish Ministry of Finance. At the time of writing, they are working as a postdoctoral research associate on the Patterns in Practice project.

Benedetta Brevini is a visiting professor at the Institute for Public Knowledge, New York University and Associate Professor of Political Economy of Communication at the University of Sydney. Before joining the academy, she worked as a journalist in Milan, New York, and London for CNBC, RAI, and the *Guardian*. She is the author of several books including *Is AI good for the Planet* (2022), *Amazon: Understanding a Global Communication Giant* (2020) and *Public Service Broadcasting Online* (2013), and is the editor of *Beyond Wikileaks* (2013), *Carbon Capitalism and Communication: Confronting Climate Crisis* (2017), and *Climate Change and the Media* (2018). She is currently working on a new volume for Polity entitled *Communication Systems, Technology and the Climate Emergency*.

Patrick Brodie is an assistant professor and Ad Astra Fellow in the School of Information and Communication Studies at University College Dublin. His research focuses on the environmental politics of digital media infrastructures, with a particular focus on dynamics of energy, extractivism, and rurality.

Teresa Cerratto-Pargman, PhD, is a Professor in Human–Computer Interaction at the Department of Computer and Systems Sciences at Stockholm University, Sweden. She conducts conceptual and empirical studies on emerging technologies and practices in the public sector, drawing from perspectives on the philosophy of technology and STS. Teresa is a research associate at the Weizenbaum Institute in Berlin, Germany. She is also a member of the Executive Committee and Associate Director for Societal Outreach at Digital Futures in Sweden.

Roderic Crooks is an assistant professor in the Department of Informatics at UC Irvine. His research examines how the use of digital technology by public institutions contributes to the minoritization of working-class communities of colour.

Nicole Dalmer is an Assistant Professor in the Department of Health, Aging and Society at McMaster University. She is also the Associate Director of the Gilbrea Centre for Studies in Aging. Nicole's current work examines the impact of social and digital infrastructures on feelings of connectedness in later life while reflecting on questions of social status and social inequality that surround and shape ageing and technology debates. She is a member of the Aging in Data Partnership Grant, funded by the Social Sciences and Humanities Research Council of Canada.

Aviva de Groot is a postdoctoral legal researcher on AI and Human Rights at the Tilburg Institute for Law, Technology, and Society of Tilburg University. She is interested in what constitutes fair and responsible knowledge practices and what that means for our approach to knowledge-generating processes and technologies (including law) and for the governance of decision practices. De Groot is involved in cross-disciplinary collaborative projects in the domains of health, administration, and AI assessment models. She came to academia with backgrounds in cabinet making, film making, legal aid and information law and is associated researcher with the African Centre for Epistemology and Philosophy of Science (ACEPS).

Catherine D'Ignazio is an Associate Professor of Urban Science and Planning in the Department of Urban Studies and Planning at MIT. She is also Director of the Data + Feminism Lab which uses data and computational methods to work towards gender and racial justice, particularly in relation to space and place.

Miriam Fahimi is a social scientist and PhD candidate in Science and Technology Studies at the Digital Age Research Center (D'ARC) at the University of Klagenfurt. Her research interests include feminist and relational approaches to technology, labour and infrastructures. She was a Marie Skłodowska-Curie Fellow in the NoBIAS – Artificial Intelligence without Bias project. Presently, Miriam is a research fellow at the Center for Advanced Internet Studies in Bochum, Germany.

Petter Falk's research is mainly focused on critical data studies and datafication of welfare services. They are currently writing a dissertation on datafication of welfare services. As a researcher, they are part of the Center for Service Research at Karlstad University and affiliated with WASP-HS (The Wallenberg AI, Autonomous Systems and Software Program – Humanities and Society). They have previously led a number of national and international innovation and development projects on data-driven systems and democracy, as a process manager and service designer at RISE (Research Institutes of Sweden). Before that, they worked as a data analyst in the private sector, and as a municipal administrator in consultation and democracy issues.

George Fletcher is professor and chair of the Database Group at Eindhoven University of Technology. He studies data systems, increasingly with a focus on people and lessening cruelty in applications of data and AI. George co-organized DataEd at SIGMOD 2023 and co-founded Social X (https://social-x.github.io/), an interdisciplinary group based in the Netherlands studying fundamental questions around technology in society.

Monika Fratczak is currently working as a teaching associate in the Department of Sociological Studies at the University of Sheffield. She completed her PhD research in 2022, focusing on emotional responses and the potential for democratic participation through data visualization about climate change in two different national contexts.

Janna Frenzel is a PhD candidate in Communication Studies at Concordia University in Tio'tia:ke (Montreal) whose doctoral research looks at interventions to reduce the environmental impacts of computing. Her work on mega data centre projects, labour and environmental organizing, and open source practices as a climate concern has appeared in *The Conversation*, *Branch Magazine*, and *Container Magazine*.

Irene Fubara-Manuel is a Senior Lecturer in Digital Media Practice at the University of Sussex. They research the colonial technoculture of border policing through digital imaging practices that mobilize black, queer, and African imaginaries.

Olga Gkotsopoulou is a PhD researcher at the Law, Science, Technology and Society Research Group at Vrije Universiteit Brussel. Her research focuses on topics of personal data protection law, transparency, and accessibility.

Lyndsay Grant is a lecturer at the School of Education, University of Bristol. Her research investigates how digital and data technologies reshape educational knowledges, practices, policies, and cultures.

Jonathan W. Y. Gray is Director of the Centre for Digital Culture and Senior Lecturer in Critical Infrastructure Studies at the Department of Digital Humanities, King's College London. He is also co-founder of the Public Data Lab and Research Associate at the Digital Methods Initiative (University of Amsterdam) and the médialab (Sciences Po, Paris). His research explores the role of digital data, methods, and infrastructures in the composition of collective life. More at jonathangray.org.

Emanuel Guay is a postdoctoral researcher in the Geography department at Concordia University and a member of the Parc-Extension Anti-Eviction Mapping Project. His research interests revolve around housing, political sociology, and ethnography.

Monika Halkort is a member of faculty of the School for Transformation at the University of Applied Arts in Vienna, Austria. From 2013 to 2020 she has been teaching and researching on data colonialism and digital infrastructures at the Lebanese American University, School for Arts and Sciences in Beirut.

Marie K. Heath is Assistant Professor of Educational Technology at Loyola University Maryland. Her research broadly asks how we might work towards more just educational futures. She investigates the intersection of education, civic engagement, and technology to foster social change. Marie is co-founder of the Civics of Technology project (www.civicsoftechnology.org/).

Arne Hintz a Reader at the Cardiff School of Journalism, Media and Culture (JOMEC), and Co-Director of the Data Justice Lab. His research focuses on the practices and conditions of digital citizenship, combining work on media activism, communications policy, and datafication. Recent co-authored books include *Digital Citizenship in a Datafied Society* (Polity, 2019) and *Data Justice* (Sage, 2022).

Rolien Hoyng is a senior lecturer in the Department of Sociology at Lancaster University, UK. Her research addresses the social and political implications of digital infrastructures and data-centric technologies in particular contexts of practice, including smart cities, waste, and ecology. Currently, she is working on the role of uncertainty in technological mediations of the climate crisis. Her fieldwork sites encompass Turkey, Hong Kong, and Western Europe. She recently co-edited a book (with Gladys Pak Lei Chong) entitled *Communication Innovation and Digital Technology: A critique of the new in a multipolar world* (Michigan State University Press).

Mergime Ibrahimi is a PhD researcher in the Ragnar Nurkse Department of Innovation and Governance at Tallinn University of Technology (TalTech), and a researcher at the FinEst Center for Smart Cities. She is a critical data scientist who studies how data are transformed in algorithmic decision-making, particularly in the context of smart mobility. Mergime's current work employs eye-tracking technology to analyse the perception of data subjects and data experts towards autonomous vehicles and explores the concept of diversity transformed in algorithmic solutions.

Fieke Jansen is a postdoctoral researcher at the University of Amsterdam and a co-principal investigator of the Critical Infrastructure Lab. Fieke's research interest is to understand how states and Big Tech are framing the relationship between environmental harm and expanding computational infrastructures.

Juliane Jarke is Professor of Digital Societies at the University of Graz. She works at the intersection of critical data studies, digital STS, and participatory design research, with a focus on digital data and algorithmic systems in the public sector, education, and for ageing populations. Juliane received her PhD in Organisation, Work and Technology from Lancaster University.

Xiufeng Jia is a Lecturer in Digital Media in the Department of Media, Journalism, and Cultural Studies at the University of Sussex. Her research areas include artificial intelligence (AI), algorithms and data in society, digital self-tracking, and digital methods. Currently, she explores responsible AI governance, design, and the everyday use of AI.

Andrea Jimenez is a Lecturer in Information Management at the University of Sheffield's Information School. Her research is centred on the role of innovation in socio-economic development, and her approach is informed by critical perspectives such as intersectional feminism, decolonial approaches, and data justice.

Anne Kaun is Professor of Media and Communication Studies at Södertörn University, Sweden. She conducts research on questions of social justice in relation to digital technologies including algorithmic decision-making systems in the public sector across domains.

Katharina Kinder-Kurlanda is a cultural anthropologist who often works in computer science. She is a Professor of Digital Culture and head of the Digital Age Research Center (D'ARC) at the University of Klagenfurt. She is interested in data, its power, politics, and practices, and in algorithms in the everyday.

Leonora Indira King is a community-based researcher and community worker in the Parc-Extension neighbourhood of Montreal. She is also the founder and coordinator of the Parc-Ex Curry Collective, a mutual aid initiative that provides catering and food distribution services in Montreal.

Leonora obtained her graduate degrees from McGill University's Department of Psychiatry, where she studied maternal health, resilience in children, and arts-based interventions for asylum seekers. Essentially, Leonora is passionate about migrant justice, bringing people together, and strengthening marginalized communities.

Rob Kitchin is a professor at the Maynooth University Social Sciences Institute and Department of Geography at Maynooth University, Ireland. He is the author or editor of 35 books, including *Data Lives* (2021), *The Data Revolution* (2022, 2nd edn), and *Researching Digital Life* (2024).

Daniel G. Krutka is a citizen, educator, and Associate Professor of Social Studies Education at the University of North Texas. His research interests concern intersections of technology, education, and democracy. He is co-founder of the Civics of Technology project (www.civicsoftec hnology.org/) and co-hosts the Visions of Education podcast (https:// visionsofed.com/).

Priya C. Kumar is an assistant professor at Pennsylvania State University's College of Information Sciences and Technology. Her research on the datafication of children's lives aims to shift digital technology discourse and design away from a focus on individual control and towards more networked understandings of privacy and agency.

Clément Le Ludec is a researcher at Telecom Paris – Polytechnic Institute of Paris, working on the Human Supply Chain Behind Smart Technologies (HUSH) project and the Digital Labor Platform project (DIPLab), exploring the impact of AI on work and especially data supply chains in French-speaking African countries.

Evan Light is an associate professor in the Bilingual Communications Programme at York University's Glendon College in Toronto, Canada. His research focuses on surveillance, technology, and policy and the uses of art in public policy education and policy making. He smokes a mean tuna.

Jakob Linaa Jensen, PhD, MA in Political Science is Associate Professor and Director of the Centre for Internet Studies at the Department of Media and Journalism Studies, School of Communication and Culture, Aarhus University, Denmark. His research is focused in the cross-field between online political participation and the algorithmic regimes of the platform economy. He has published four monographs, three edited volumes, and more than 30 international journal articles. His most recent book is *The Medieval Internet* (Emerald Publishing, 2020).

Stine Lomborg is a professor at the Department of Communication at the University of Copenhagen, Denmark, where she directs the Centre for Tracking and Society. She has published extensively on diverse people's everyday engagement with digital tracking, datafication, and automated decision-making systems across personal, work, and institutional contexts.

Giovanna Mascheroni is a Full Professor in the Department of Communication, Università Cattolica del Sacro Cuore. She is the leader of WP6 in the H2020 project ySKILLS, and the PI of DataChildFutures, a project investigating the data practices of Italian families with children aged 0–8 years old. Her work focuses on the social consequences of digital media, datafication, and AI for children and young people.

Ellouise McGeachey is a PhD candidate in the joint Communications and Culture Programme between York and Toronto Metropolitan University. Her research focuses on genetic surveillance, necropolitics, and privacy policy. Ellouise has ten years' professional experience in the private sector. She contributes both academic and real-world expertise to this chapter.

Dan McQuillan is a lecturer in Creative and Social Computing at the Department of Computing at Goldsmiths, University of London. Prior to academia, he worked as Amnesty International's Director of E-communications. His research focuses on the resonances between forms of computational operation and their specific social consequences, especially in relation to machine learning and AI.

Itzelle Medina Perea a Lecturer in Information Systems at the Information School, The University of Sheffield. She is interested in exploring the factors that shape data flows, data practices, and the uses of data and information systems. Itzelle has experience beyond academia, having worked for the World Wide Web Foundation (A4AI Good Practices Database – Alliance for Affordable Internet), and for the National Council of Science and Technology (CONAHCYT Mexico).

Alex Megelas is the Manager of Research Innovation and Business Development at Concordia University's Applied AI Institute, and an affiliate facilitator of the Centre for Community Organizations (COCo). He teaches at the Saint Paul University School of Conflict Studies.

Iris Muis, University of Utrecht, is team leader of the Data School – a research and teaching platform that investigates the impact of technology on society. Iris specializes in data ethics and its implications for public management. She is the co-author of *DEDA: the Data Ethics Decision*

Aid, and *FRAIA: the Fundamental Rights and Algorithms Impact Assessment*. Both instruments are household names among public servants working in technology and innovation, in the Netherlands and abroad.

Denis Newman-Griffis (they/them) is a Lecturer in Data Science in the University of Sheffield Information School. Their work investigates the practices at the heart of responsible use of AI and machine learning, with particular focus on healthcare, disability, and the scientific research process. Denis' research ranges from development of natural language processing technologies for health data to critical analysis of AI design from a disability lens. They are a Visiting Fellow of the National Institutes of Health Clinical Center, Research Fellow of the Research on Research Institute, and lead multiple responsible AI projects funded by the Research on Research Institute and the UK's Arts and Humanities Research Council.

Eric Nost is an Assistant Professor of Geography at the University of Guelph and a member of the Environmental Data and Governance Initiative. His interests lie in understanding how environmental data systems come to be, as well as experimenting with them towards more just and equitable ends.

Jonathan A. Obar is an associate professor in the Department of Communication and Media Studies at York University. His research addresses relationships between information policy, digital technologies, and civil liberties. Recent publications address connections between corporate transparency, online consent, and information protections.

Susan Oman, University of Sheffield, researches how data and evidence work from the perspectives of different stakeholders and 'follow the data' around, in order to understand data contexts 'in the round'. They often work in and with public-sector organizations, writing recommendations on particular policy problems. At the time of writing, they are a research-policy fellow based in the UK Government's Department for Digital, Culture, Media and Sport and they are also a Lecturer in Data, AI and Society at the University of Sheffield, where they lead a module called Data Science Foundations and Contexts.

Luci Pangrazio is a senior lecturer at Deakin University, where she is based in the Centre for Research for Educational Impact (REDI). Luci is an expert in datafication, young people's digital and data literacies, and digital cultures. She is currently Chief Investigator in the Australian Research Council funded 'Digital Child' Centre of Excellence. Recent books include: *Young People's Literacies in the Digital Age* (Routledge, 2018) and *Critical Data Literacies* (MIT Press, 2023, with Neil Selwyn).

Lucy Pei is a PhD candidate at the University of California, Irvine, in the Department of Informatics. Her research broadly examines tech-for-good endeavours across different contexts. Her dissertation is focused on transnational labour activism among digital platform workers.

Nikolaus Poechhacker is a postdoctoral researcher at the Digital Age Research Center (D'ARC), University of Klagenfurt. In his work he researches the relationship between democratic institutions, law, social order, and algorithmic systems in various domains, bringing together perspectives from media theory, STS, computer science, and sociology.

Lindsay Poirier is a cultural anthropologist of data advocacy, governance, and infrastructure. Interlacing methods in cultural analysis and exploratory data analysis, Poirier critically examines how meaning gets made from data – by whom, for whom, under what conditions and towards what ends. She is an Assistant Professor of Statistical and Data Sciences at Smith College.

Ana Pop Stefanija is a PhD researcher at imec-SMIT, Vrije Universiteit Brussel. Her research is focused on the question of embedding human agency in algorithmic and automated decision-making systems, while accounting for the complex interplay of people, artefacts, and practices.

Paul Quinn is a research professor at the Law, Science, Technology and Society (LSTS) Research Group at the Vrije Universiteit Brussel. His research interests include the stigmatization of vulnerable groups and the delivery of healthcare, and his research covers topics such as data protection, privacy issues, and problems related to stigmatization and discrimination. He is part of the Health and Aging Law Lab (HALL) at LSTS where he coordinates research on such issues.

Joanna Redden investigates the social justice implications of the datafication of governance. This work involves a combined approach that includes researching where and how government bodies are using algorithmic and automated decision-making systems, particularly in the area of social services, as well as investigating the impacts of changing systems on ways of knowing, decision-making, service delivery, service users, and people targeted by these systems.

Elise Renkema holds a Master of Science in Research in Public Administration and Organizational Science (2023, Utrecht University) and specializes in algorithmic accountability in local governance. She worked at the Data School (Utrecht University, the Netherlands) and was part of multiple projects surrounding the impact of technology on national, regional, and local government. She is co-author of the *Digitale Raadgever* (Digital Councillor), a tool to help municipal council members make responsible decisions about data and AI projects. She currently works as a municipal council adviser in the Netherlands.

Alessandra Renzi is Associate Professor of Critical Media Production in the Department of Communication Studies, Concordia University, Montreal, where her research-creation explores the intersection of media, art, and civic engagement. Alessandra's latest book is *Hacked Transmissions: Technology and connective activism in Italy* (University of Minnesota Press, 2020).

Theo Röhle is interested in critical perspectives on datafication, at the intersection of science and technology studies, digital media studies, and historical media studies. He is Associate Professor in Media and Communication Studies at the Department of Journalism, Media and Communication, University of Gothenburg, Sweden.

Julia Rone is a postdoctoral researcher at the Minderoo Centre for Technology and Democracy at CRASSH, Cambridge. She was previously a Wiener-Anspach postdoc at the Université libre de Bruxelles and the Department of Politics and International Studies (POLIS), Cambridge. Julia has a PhD from the European University Institute in Florence with a thesis on mobilizations against free trade agreements. Her current research focuses on the democratization of Internet policy regulation. She has written on hacktivism, digital disobedience, and more recently, the rise of far right media in Europe.

Akrati Saxena is an assistant professor in the Computational Network Science Group at the Leiden Institute of Advanced Computer Science (LIACS), Leiden University, the Netherlands. She has worked extensively on network science and data science. Currently, she is working on algorithmic fairness in Network Science and Data Science, which focuses on analysing biases in existing systems, designing fair algorithms, defining fairness constraints and fairness-aware evaluation metrics, and modelling affirmative actions.

Neil Selwyn is a professor at the Faculty of Education at Monash University. Recent books include: *Should Robots Replace Teachers? AI and the future of education* (Polity, 2019), *Critical Data Literacies* (MIT Press, 2023, with Luci Pangrazio), and the third edition of *Education and Technology: key issues and debates* (Bloomsbury, 2021).

Alexander Serebrenik is Professor of Social Software Engineering at the Eindhoven University of Technology, the Netherlands. His research goal is to facilitate the evolution of software by taking into account social aspects of software development. His work tends to involve theories and methods both from within computer science (for example, theory of socio-technical coordination; methods from natural language processing, machine learning) and from outside of computer science (for example, organizational psychology). The underlying idea of his work is that of empiricism, that is, that addressing software engineering challenges should be grounded in observation and experimentation, and requires a combination of the social and the technical perspectives. Alexander has co-authored the book *Evolving Software Systems* (Springer Verlag, 2014), the upcoming volume *Equity, Diversity and Inclusion* (APress, 2024) and more than 240 scientific papers and articles. He is actively involved in the organization of scientific conferences and is a member of the editorial board of several journals. He has won multiple best paper and distinguished reviewer awards.

Leslie Regan Shade is a professor at the Faculty of Information at the University of Toronto. Her research focus since the mid-1990s is concerned with the social and policy aspects of information and communication technologies (ICTs), with a focus on gender, youth, and political economy. Her recent work is supported by the Social Sciences and Humanities Research Council of Canada partnership grant, The eQuality Project.

Sepideh Shahamati is a PhD candidate in the Department of Geography, Planning, and Environment at Concordia University and a member of the Parc-Extension Anti-Eviction Mapping Group. Her research revolves around the intersection of gentrification, cartography, and storytelling. In her current work, she is exploring the use of stories and maps for representing the impacts of forced displacement in Montreal.

Tamar Sharon is a philosopher of technology who studies how digitalization destabilizes public values and how best to protect them. She is Chair of the Department of Ethics and Political Philosophy and Co-Director of the Interdisciplinary Hub for Digitalization and Society (iHub) at Radboud University in the Netherlands.

Karen Louise Smith is an associate professor in the Department of Communication, Popular Culture and Film at Brock University. Her recent research work on datafied childhood was supported by The eQuality Project from the Social Sciences and Humanities Research Council of Canada and the Social Justice Research Institute internally at Brock.

Marthe Stevens studies the ethical and societal consequences of new technologies, primarily in health and medicine and in education. She is an assistant professor at the Department of Ethics and Political Philosophy Interdisciplinary Hub on Digitalization and Society (iHub) at Radboud University in the Netherlands.

Linnet Taylor is Professor of International Data Governance at Tilburg Law School in the Netherlands. Her research focuses on processes of integrating social justice concerns with the shaping and regulation of digital technologies. She has an interdisciplinary background including development studies and migration studies, and has previously worked at the Oxford Internet Institute and the University of Amsterdam.

Marlee Tichenor, University of Edinburgh, is trained as a medical anthropologist, and as she has expanded into the field of critical data studies, she has analysed how social data are produced, processed, and used by various actors who work with the stated goal of social justice and equity. These data have included malaria rapid diagnostic test results and the monitoring of the delivery of long-lasting insecticidal nets (LLINs) in Senegal, as well as the proportion of household income spent on healthcare. She is engaged in providing critical pedagogies for data practitioners, including recently developing and teaching Critical Perspectives in Data Science to prospective data scientists at Durham University.

Bernard van Gastel is Assistant Professor of Sustainable Digitalization in Computing Science at Radboud University. Besides his research in predicting the energy consumption of software, he researches sustainability in a broader sense. This includes the effects of software on society, such as inclusion, equal opportunities and privacy.

Gijs van Maanen is an assistant professor at the Tilburg Institute for Law, Technology, and Society (TILT). His research interests range from data governance, political theory, and political economy, to STS.

Hunter Vaughan is senior research associate at the Minderoo Centre for Technology and Democracy, University of Cambridge. Dr Vaughan is author or co-editor of numerous articles and books, including *Hollywood's Dirtiest Secret: the hidden environmental costs of the movies* (Columbia University Press, 2019), the *Anthem Handbook of Screen Theory* (2018), and *Film and Television Production in the Age of Climate Crisis* (2022), and is a founding editor of the *Journal of Environmental Media*. He is co-director of the Global Green Media Network and co-Principal Investigator on the Sustainable Subsea Networks project.

Tamara Vukov has been teaching at the Université de Montréal since 2013. She holds a PhD from Concordia University, and from 2011 to 2013 worked as a postdoctoral researcher at the Center for Mobilities Research and Policy in the Department of Culture and Communication at Drexel University (Philadelphia, USA) on an SSHRC fellowship.

Quinn White is interested in the ethical implications of how we use data in medical research. Her interest was sparked while studying under Lindsay Poirier as part of the Statistical and Data Sciences department of Smith College, and she carries this motivation forward as a PhD student at the University of Washington.

Lorenzo Giuseppe Zaffaroni is a postdoctoral researcher in the Department of Communication, Università Cattolica del Sacro Cuore. His work focuses on fields of cultural production, digital media, and datafication. He is currently part of the DataChildFutures research project, where he focuses on the interplay between socio-material practices and datafied environments with a mixed-method approach.

Irina Zakharova is a postdoctoral researcher at the Leibniz University Hannover. Research presented in this book was conducted when she was affiliated with the University of Bremen. She studies datafication of welfare and public schooling with particular focus on data, care, work, and infrastructures required for organizing these domains from an interdisciplinary perspective of STS, digital media studies, and sociology.

Dialogues in Data Power: Shifting Response-abilities in a Datafied World

Juliane Jarke and Jo Bates

Introduction

How and what responsibilities are shifting in a datafied world? Who is doing the shifting and how do diverse groups of people experience and respond to these shifts? How can we realize 'response-ability' (Haraway, 2016) in research and beyond as a sense of shared agency and mutual responsiveness that goes beyond individual interests, encompassing the well-being of communities and more-than-human worlds? This book responds to these questions through a series of dialogical chapters between scholars in critical data studies (for example, Kitchin and Lauriault, 2014; Iliadis and Russo, 2016; Dalton et al, 2016; Hepp et al, 2022). We have come together across disciplines, geographies, and modes of engaging with – what we call – *data power* to explore, discuss, and challenge pressing issues and emerging themes in our growing interdisciplinary endeavour to make sense of and interfere with processes, imaginaries, and effects of datafication.

The book's title is Dialogues *in* Data Power and not *about*, as we take our own positionality and entanglements in the ever increasing datafied world as a starting point. The book itself is an experiment in facilitating interdisciplinary dialogue and collective scholarship among 80 researchers through nine collectively authored chapters. Contributors to each chapter were invited based on their presentations at the 4th Data Power Conference which attracted 175 participants and took place in June 2022 simultaneously in Canada, Germany, the UK, and online.[1] In a series of workshops for each of the chapters, the invited contributors explored their various perspectives, experiences, and responses to the chapter's theme. They then collectively wrote a joint introduction to the chapter, as well as individual sections that provide their own perspective. This responsive process led to what authors in different chapters refer to as a 'kaleidoscope' or 'braid' of scholarly engagement. The process, hence, did not aim to provide a conclusive view on any given theme but allowed for a *diffractive writing* of multiple and also differing or disagreeing perspectives. To further increase the dialogue, we also invited scholars – mostly from outside the Data Power realm – as discussants for the individual chapters. Their task was to provide a response and situate the arguments in their own research. Through this form, we have created a space for dialogue and mutual encounter that is difficult to find otherwise in such an interdisciplinary field. We do not have any formal association, there are only a few conferences and academic publications that attract this wide variety of academics working on datafication in such diverse disciplines as media and communication studies, information studies, STS, information systems, software engineering, data science, law, education, age studies, cultural anthropology, critical geography, and gender studies. At the end of this book, we present and reflect on our experiences with this kind of scholarly writing process.

Chapter overview

Overall, this volume encompasses nine collectively written chapters which we summarize below. We then proceed with a *diffractive reading* across the nine chapters and identify five cross-cutting themes relating to shifting response-abilities in a datafied world.

1. Configuring Data Subjects. The first chapter examines how individuals become subjects of datafication within data-driven systems, emphasizing the importance of historical, social and power dynamics. It questions data objectivity by exploring different domains in which individuals come to be configured as data subjects, such as self-tracking, ageing, disability, and autonomous vehicles.

2. Children as Data Subjects: Families, Schools, and Everyday Lives. The second chapter examines the increasing digitized surveillance of children's lives in three contexts: 'sharenting' in the USA, Italian family privacy practices, and English secondary schools. It explores themes of neoliberalism, subjectification, risk, and decision-making. The chapter advocates for a data justice approach to counter the datafication of childhood, emphasizing the need for governmental regulation and inclusive decision-making involving children's voices.

3. In/visibilities in Data Studies: Methods, Tools, and Interventions. This chapter examines research practices of rendering data visible and the inherent power dynamics and imbalances that accompany them. The authors have developed a dialogue in three interconnected parts, each focusing on the intersection of in/visibility and power: (1) the challenges of producing knowledge about and with data; (2) empirical case studies analysing the consequences, methodological opportunities, and challenges of data visibility;

and (3) tool-based interventions aiming to bring alternative data framing and narratives to the fore.

4. *People's Practices in the Face of Data Power.* The authors of this chapter present and discuss changing power dynamics between state, citizens, and industry, with individuals having little control over data collection and algorithmically informed decisions that govern their lives. The chapter explores emerging power imbalances and points to possibilities of agency through practices of consent and refusal, to citizen participation, and other forms of data activism. In doing so, the authors contribute to reshaping data power 'from the bottom up' and propose a people-centred and radically contextualized approach for our data futures.

5. *Practitioner Interventions in Data Power.* This chapter explores the different identities and responsibilities within the data ecosystem, from data scientists to policy makers, highlighting tensions and reflections on the work of these different data practitioners. Authors explore what data work looks like in practice and through which ethical, legal, and societal considerations practitioners orient their work.

6. *Critiques of Data Colonialism.* The authors in this chapter engage in a critical intervention of Couldry and Mejias' (2020) concept of 'data colonialism'. Bringing contrasting perspectives to the discussion, the authors put forward arguments that offer fundamental critiques of the data colonialism concept, as well as examining how it might be extended to address issues of environmental injustice and historicized into the power relations of medieval feudalism. Together they problematize a simple reading and application of the 'data colonialism' concept for understanding contemporary data power.

7. *Environmental Data Power.* This chapter examines the relationship between data and the environment, through an exploration of the practice of environmental sensing, data-driven representations of climate change, and the environmental impacts of data centres. The authors ask: how does 'environmental data power' as a concept that encompasses all of these practices and contexts operate across various registers? In bringing together varied cases based on their independent research, the authors add to the growing conversation about data and environmental justice.

8. *Data and Technological Spatial Politics.* This chapter brings together a group of authors that have been working on independent research projects examining the Montreal neighbourhood Parc-Ex, where local activists oppose processes of gentrification and displacement driven by an emerging AI ecosystem meant to boost Canada's innovation and platform economies, with other researchers exploring the spatial politics of data and technology at a global scale. Through their combination of macro and micro analysis of the spatial politics of data and technology, they identify three forms of data power that help them understand agency within these spaces: hegemonic data power, powering data, and data counter-power.

9. *A Canon Is a Blunt Force Instrument: Data Science, Canons, and Generative Frictions*. This chapter is collectively authored by a group of academics from three Dutch universities and from a diverse set of disciplines. It emerged from a keynote dialogue at the 4th Data Power Conference. The authors take their different disciplinary backgrounds and research fields as starting points to ask how different conceptions of the same term – here AI and trust – can be generative or problematic for critical engagements with data-driven technologies.

Cross-cutting concerns: shifting response-ability

Each of the individual chapters hence provides an interdisciplinary engagement with different emerging and important themes in critical data studies. We now turn to five cross-cutting concerns about *shifting response-abilities in a datafied world* that we identified across the chapters: (1) the responsibilization of individuals and communities through processes of datafication, (2) the abilities of individuals and communities to respond to data power, (3) the endeavour to design responsible data-driven systems, (4) questions around what responsible data studies research may look like, and (5) how we as critical data studies scholars can become response-able to each other.

First, contributions in this volume articulate concerns about the *responsibilization of individuals and communities* who are subject to an increasing datafication. Here responsibility is shifted to the individuals and/ or communities who experience data power. For example, in Chapter 1, 'Configuring Data Subjects', Nicole Dalmer considers the responsibilization of older adults through data-driven technologies. Older persons self-track their lives and bodies to become more self-knowledgeable, they simultaneously also become more responsible for their well-being in a public discourse that more often than not positions older populations as a burden and threat to our healthcare and social systems. The authors of Chapter 2, 'Children as Data Subjects', discuss the responsibilization of guardians vis-à-vis an increasing datafication of children's lives. Here responsibilities are assigned to parents or teachers as 'responsible mediators' to ensure children's safety and privacy.

A second and related concern across the chapters is the *ability of individuals and communities to respond to data power*. In Chapter 2, authors consider for example a 'data justice' approach as a 'potential resistance and response to the datafication of childhood'. The authors in Chapter 4, 'People's Practices in the Face of Data Power', call the ability to respond 'collective agency'. Examples that authors list include community data activism, modes of participatory governance, and citizen participation. This requires understanding and supporting the conditions under which individuals and collectives are able (or enabled) to respond to data power. Here collective response-ability refers to the ability to refuse, resist, and challenge regimes of data power. Authors explore what kinds of resources are required to do so, where to find them, or how to create them collectively. Catherine D'Ignazio adds to the authors' contributions in her response the heterogeneous activist practices that do not respond against data power, but use data science as an 'imperfect informatic tactic in a deeply asymmetrical environment'. It is a practice that is part of a broader 'constellation of efforts that are working towards the restoration of rights, the healing of communities, and the longer-term work of structural transformation'. Dan McQuillan summarizes this similarly in his response: 'Resistance is more than collective refusal, it is a commitment to possible alternatives.' In Chapter 8, 'Data and Technological Spatial Politics', several contributors explore community activism through a case study of Montreal, where local activists oppose processes of gentrification and displacement driven by an emerging AI ecosystem meant to boost Canada's innovation and platform economies.

Third, responsibility has become a – to some extent token – *promise for designing and implementing data systems.* Phrases such as 'responsible AI' or 'fairness, accountability and transparency' point to the increasing awareness of the industry but also data science practitioners about their responsibility in producing systems that do not inflict harm on individuals, communities, and the environment. In Chapter 5, 'Practitioner Interventions in Data Power', the authors consider the responsibilities and resulting interventions of professionals engaged with data work – from data scientists (in training) to civil servants, journalists, and activists. Authors ask: Who, why and how do practitioners working with data as part of their professional practice feel responsible? To whom/what do they feel responsible? In Chapter 9, 'A Canon Is a Blunt Force Instrument', the authors consider their educational responsibilities in teaching data science and related subjects. In addition they explore differently assigned responsibilities for the design and regulation of data-driven technologies. Concerns related to environmentally responsible design are addressed in Chapter 7, 'Environmental Data Power', but also in Chapter 6, 'Critiques of Data Colonialism', in which Benedetta Brevini critiques the concept of data colonialism in relation to environmental justice. Collectively, authors examine the various environmental harms that may arise as a result of the design of digital infrastructure components such as data centres and submarine cables, as well as observing the energy required to power AI models. They call for a better understanding of these forms of 'energetic data power' when considering the infrastructural design, and introduce environmental justice critiques that aim to address some of these issues.

A fourth way in which shifting response-abilities are important to authors, are the ways in which *we – as researchers – engage responsibly with our research participants and are response-able to what we witness and learn*. Donna Haraway (2016) proposes a feminist ethics of response-ability as attending to affect, entanglement, and rupture. With Haraway, critical data studies scholars can ask how we may 'cultivate the capacity of response-ability' (p 35). In particular in Chapter 3 'In/visibilities in Data Studies', authors consider questions of methodology and engagement in critical data studies. They forefront the question of how our methods, tools, and interventions produce visibilities and invisibilities of and within data power. Their guiding questions include: 'How do we chart the contours of visibility for something as ephemeral and capricious as data? How do we evaluate what we can see – its truth, its accessibility and its social value? How do we make sense of when and for whom data visibility is productive?' Chapter 6, 'Critiques of Data Colonialism', is exemplary of scholarship that grapples with questions of how research responds adequately to the lived experiences of historic and contemporary colonized populations. For example, Clément Le Ludec provides insights into a case study of low-paid data workers in Madagascar, while the chapter's discussant Andrea Jimenez argues that the ambiguity of the term data colonialism is not just a 'conceptual problem' but has 'material implications'. Here Haraway (2016) can provide guidance when she urges researchers to consider the response-abilities that arise from their practice and calls on them to 'stay with the trouble'. This can lead to 'viral response-abilities' (p 114). Research methods are contagious and mutate, and researchers in critical data studies may want to consider how we can spread methods for (more) critical engagement and careful intervention (Springgay and Truman, 2018). Ultimately this points to our response-ability for configuring the apparatus of research (Barad, 2007) and struggles over what research will count as a legitimate account of datafied realities (Haraway, 1988).

A fifth way in which response-ability plays out in this book, is *how we become response-able to each other*. Chapter 9, 'A Canon Is a Blunt Force Instrument', demonstrates one such endeavour to enable generative responses that grew out of the keynote panel of the 4th Data Power Conference. The authors are all based at Dutch universities, but are part of very different disciplines – spanning from law and philosophy to software engineering and data science. In their contribution they state: 'We talk, and sometimes even discuss our research with one another. But do we also understand each other? What would that even mean? When we talk about "data", do we talk about the same thing? Is that even necessary?' In their chapter, the authors explore and juxtapose 'canonical objects' in their teaching such as 'artificial intelligence' and 'trust', and consider the role of such objects for dialogue across disciplines. The chapter is a showcase for how the collective 'we' of critical data studies scholars can become response-able to each other and the wider communities that intersect with our work. Our book is hence an attempt to respond to each other and our research in a different form and process. In particular in an interdisciplinary and emerging field like critical data studies, where professional associations are non-existent, degree programmes are only emerging, and conferences such as Data Power or Data Justice are still relatively few, we need spaces for dialogic encounters. The process of writing this book and coming together in our chapters, enabled us to follow an approach to 'slow science' (Stengers, 2018) in which we carefully explored each other's standpoints, experiences, and perspectives in order to find a common ground from which we can embark to not only critique but to shift data power.

For all five concerns, authors emphasize the importance of *historically developed and manifested power inequalities*. For example, authors in Chapter 5, 'Practitioner Interventions in Data Power', consider how 'contemporary uses of data are built upon longer histories of unjust, capitalist, and colonialist ways of knowing and controlling'. Denis Newman-Griffis argues in Chapter 1 that data have always both a history and a purpose. The chapters also demonstrate that part of our ability to collectively respond to data power may be based on knowledges and experiences of the past. In particular, authors in Chapter 6, 'Critiques of Data Colonialism', situate current data harms inflicted on marginalized and minoritized groups, in historical injustices, power inequalities, and extractivist logics.

So, what might we learn from our diffractive reading of the nine chapters of this book? We have demonstrated that authors raise concerns about who or what is doing the shifting of responsibilities but also attend to different kinds of abilities to respond in the face of data power. Overall, this points to the ways in which data power and different forms of response-ability are intricately linked. For critical data studies this leads to a set of new and potentially guiding questions for our research and educational practices: How can we enable ourselves and others to respond to the challenges and potential harms of datafied societies? What forms can and should our response take? For example, what responsible data practices may practitioners need to develop? What forms of collective response may individuals and community organizers develop in the face of data power? How well are our legal systems equipped to respond and regulate our increasingly datafied world in the interest of those who become data subjects and are impacted by data-driven decision-making? We believe that answering these questions can only be a collective endeavour.

Ultimately *critical* research requires the cultivation of 'response-ability' which is more than critique, because 'it requires the risk of being from some worlds rather than others and helping to compose those worlds with others' (Haraway, 2016: 178). We believe that the chapters of this book demonstrate that after more than a decade of critiquing 'big data', 'datafication', and more recently 'artificial intelligence', the field of critical data studies has moved forward to also formulate alternative visions of more just and sustainable datafied futures. Critique in data studies has not 'run out of steam' (Latour, 2004), but as this book demonstrates, found new abilities to respond and assume response-ability.

For now, we entrust the reader to explore the various threads of the individual chapters and their contributions as an engagement with an ongoing dialogue. Before we close the introduction and to facilitate the reading, we give a brief summary of 'how to read the book'.

How to read the book

Each of the collectively written chapters consists of an introduction that was jointly written by the author teams. This introduction sets the theme, a provocation, or the main question of the chapter as well as an overview of the chapter's structure. Each chapter then consists of four to seven sections in which individual authors or author teams explore the theme, provocation, or main question from their perspective. Collective writing does not require consensus and is not based on an individual author but is an emergent process that allows exploration of a multiplicity of perspectives, experiences and positionalities to facilitate dialogue (Jandrić et al, 2023). The individual sections are hence complementary and may present diverging perspectives. To summarize their dialogical encounter, each chapter collective provides a brief conclusion with what they consider to be the main points. The conclusion is followed by one or two responses of discussants who situate the arguments in their own research and complement the perspectives presented. Depending on the number of chapter authors, we either invited one or two discussants.

Hence, in line with the book's title *Dialogues in Data Power*, the format facilitates dialogue between the collective authors and discussants, and makes these dialogues visible for the reader. We believe that this will strengthen the overall endeavour of the field of critical data studies and demonstrate an inclusive and supportive innovation in academic practice as this dialogic format introduces readers to succinct arguments within critical data studies while also appreciating the breadth of interdisciplinary perspectives across the field.

Note
1   See also www.datapowerconference.org

References

Barad, K. (2007). *Meeting the Universe Halfway: Quantum physics and the entanglement of matter and meaning.* Duke University Press.

Couldry, N., and Mejias, U. A. (2020). *The Costs of Connection: How data is colonizing human life and appropriating it for capitalism.* Stanford University Press.

Dalton, C. M., Taylor, L., and Thatcher, J. (2016). Critical data studies: A dialog on data and space. *Big Data & Society*, 3(1).

Haraway, D. (1988). Situated knowledges: The science question in feminism and the privilege of partial perspective. *Feminist Studies*, 14(3), 575–599.

Haraway, D. (2016). *Staying With the Trouble: Making kin in the Chthulucene*. Duke University Press.

Hepp, A., Jarke, J., and Kramp, L. (eds) (2022). *New Perspectives in Critical Data Studies: The ambivalences of data power*. Springer.

Iliadis, A., and Russo, F. (2016). Critical data studies: An introduction. *Big Data & Society*, 3(2).

Jandrić, P., Luke, T. W., Sturm, S., McLaren, P., Jackson, L., MacKenzie, A., et al (2023). Collective writing: The continuous struggle for meaning-making. In *Postdigital Research: Genealogies, Challenges, and Future Perspectives* (pp 249–293). Springer Nature Switzerland.

Kitchin, R., and Lauriault, T. (2014). Towards critical data studies: Charting and unpacking data assemblages and their work.

Latour, B. (2004). Why has critique run out of steam? From matters of fact to matters of concern. *Critical Inquiry*, 30(2), 225–248.

Springgay, S., and Truman, S. E. (2018). On the need for methods beyond proceduralism: Speculative middles, (in)tensions, and response-ability in research. *Qualitative Inquiry*, 24(3), 203–214. https://doi.org/10.1177/1077800417704464

Stengers, I. (2018). *Another Science Is Possible: A manifesto for slow science*. John Wiley & Sons.

1

Configuring Data Subjects

Nicole Dalmer, Denis Newman-Griffis, Mergime Ibrahimi, Xiufeng Jia, Doris Allhutter,[1] Katrin Amelang,[2] and Juliane Jarke[3]

Introduction

The collection and use of data about individual people has shifted dramatically over the last century, from a specialized practice to a near-universal norm. Once the domain of census-takers and actuaries, data about ourselves and our behaviours are now expected in nearly all situations, and we often enthusiastically collect such data about ourselves (Neff and Nafus, 2016). As data collection has changed, so too has data use: larger and richer data sets have co-evolved with techniques to analyse them and data analysis has become a part of everyday life, from the trains we ride, to the coffee we buy, to the ways government policies are designed and evaluated (LaValle et al, 2010).

But who decides what 'counts' as data, or how they are collected? And what impact do these choices have on us and the world we live in? When data are everywhere, it is easy to lose sight of the fact that they come from somewhere, and that the data we have are not the only data that could be. The ubiquity of data collection and analysis magnifies the impact of how data are conceptualized exponentially; at the same time, the increased sophistication and commercialization of data technologies have made these questions of conceptualization and design increasingly opaque. As data become ever richer, more detailed, and more incomprehensible (for example, Martin's (2019) investigation of her multi-gigabyte personal data profile from Google), there is an increasing perception that data are not just objective but perhaps even complete – that is, that given enough sensors, apps, and hard disks, we can represent an individual in toto (see Braun's (2021) discussion of this perception in the emerging science of digital twins).

This chapter challenges this perception, and brings to light the often overlooked questions of how quantification of people in data occurs and the societal and individual implications of the configuration of people as data subjects. To illustrate the multiplicity of datafication, we intentionally avoid a single definition of data subjects, allowing each contributing author to reflect individually on their interpretation of the term and its implications. Using specific examples of datafication to inform our analysis, we examine the process of datafication writ large, and what it looks like, to bring critical methods to bear on datafication and its impacts. Each section in this chapter engages with key questions in the interdisciplinary nature of how individuals are configured as data subjects in the datafication process and the methodological tools that can be brought to bear on analysing datafication, as well as important gaps in these tools that contemporary data processes highlight.

In addition to the configuration of individuals as data subjects, we put forth initial reflections on the ways in which datafication is used both to construct and to represent population groups and collective identities. While the politics of group formation and representation in data present distinct questions beyond the scope of this chapter in terms of data as a site for political enquiry and action (Beraldo and Milan, 2019), we highlight the importance of distinguishing between the datafication of populations and the datafication of individuals within those populations. Similarly, we draw out the distinctions between representing an individual's personhood and their actions, networks, or other outward spurs. There are invaluable questions shared between these scenarios – who is collecting what data, what power relationships are involved, and so on – but there are also fundamental differences between using data to represent and/or learn about the world, which population data generally aim to convey (Mooney et al, 2015), and using data to represent and/or learn about an individual (our focus in this chapter). These processes are highly interrelated, but require different methodologies as our ways of analysing data configurations continue to develop.

Examining the configuration of data subjects is both a timely and an often overlooked need. In addition to debates about digital twinning and ever-greater personalization of healthcare (Armstrong, 2017) and marketing and sales (McFall et al, 2020), among others, the ways in which people are, or fail to be, represented in data directly inform contemporary debates about responsible and ethical use of artificial intelligence (AI) (Sambasivan et al, 2021; Werder et al, 2022). Each of the following subsections further draws out the cross-disciplinary interweaving of dialogues around datafication, in disability data science, the quantified self, population ageing, and mobility. As data and datafication are increasingly implicated in all areas of the 'data-driven' society, these provide instructive examples for investigating other connections and questions in new and emerging areas.

In our writing of this chapter, we seek to equip readers with the tools to begin asking and answering questions about who becomes datafied, how this process is realized, and how datafication interacts with broader, historical debates about personhood and the self.

Design decisions and the history of data – Denis Newman-Griffis

Denis identifies as a White, non-binary, neurodivergent academic. They draw on their work on methodologies in natural language processing and AI, data science, and critical disability studies.

One of the first questions an informed data consumer should always ask when encountering a new data set is: Where did this come from? The companion question, which is often left unnoticed and unasked in the shadow of the enticing realms of data use, is: How did this come to be this way?

Data have both history and purpose. Measurements are taken by certain people for certain reasons; answers are elicited to specific questions from specific respondents. The purposes for which data are collected may be malicious, beneficial, or banal; sample populations may be representative or biased; data collectors may recognize and be informed by the situated nature of their work or perform it merely by rote. Each of these characteristics informs the history of data and affects its representational power and impact on the world.

More than this, however, data have conceptual history. The process of narrowing down the infinite space of information we could represent about a person to the finite subset we choose is a process of decisions, taken consciously or unconsciously over and over again each time we work with a person as data. We can imagine a data funnel representing the sequence of decisions to turn a desire for information into data that approximates that information (Figure 1.1). The act of datafication involves reframing an infinitely complex human being as a data subject, to be represented via a set of finite data. A person's lived complexity is thus approximated via a clumsy phantasm of data, and it is the decisions represented in the data funnel that configure the shape of this data ghost. Vitally, these decisions also provide anchor points for deconstructing this configuration – and for contesting it.

Consider the case of a person applying for government benefit support on the basis of disability. From an 'equal treatment' policy perspective (Mabbett, 2005), as well as from a decision consistency and accountability view (Gallicchio and Bye, 1981), the decision is typically made on the basis of standardized criteria using a fixed set of information sources to represent the person's unique disability experience. Already the process prefers information that can be directly elicited or measured and represented in codifiable ways – ineffable and subjective aspects of a person's perceptions or experience are disfavoured and likely not captured (Osterweis et al, 1987;

Figure 1.1: The data funnel: conceptual illustration of the datafication process

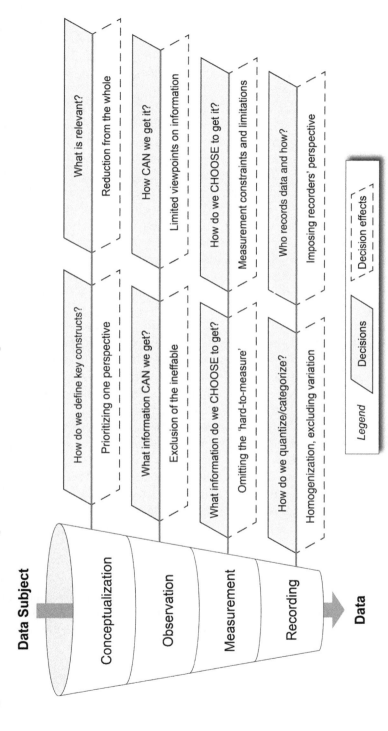

Note: Conceptual illustration of the datafication process by which an infinitely complex individual person, framed as a 'data subject', becomes represented in finite data. Example questions illustrating design and operational decisions at each stage of the process are shown along with some of their effects on what the data represent.

Newman-Griffis et al, 2022). Identifying what specific information to elicit, then, requires deciding: How do we choose to define disability, and what is relevant to that definition? A political and interpersonal definition necessitates different types of information from the context-focused social model, and yet different information from still-frequent medicalized views (Newman-Griffis et al, 2023).

These decisions continue down the data funnel. While it would simplify analysis, there are no single attributions of agency or motive to be made in this process: a state may require collecting information on medical conditions for statistical as well as decision-making purposes (Harrison et al, 2021); benefits agents may seek or avoid additional information within the narrow bounds afforded to them, to help an applicant or make a decision faster (Barth et al, 2017); and applicants may choose to emphasize, omit, or reframe certain information to better align with policy requirements (Halpern and Hausman, 1986). Nonetheless, power dynamics in the data collection process are clearly weighted in favour of those higher up the decision chain: a benefits applicant can do very little to answer questions that are not asked, and their answers may be omitted from recorded data as 'irrelevant' regardless.

These conceptualizations of information and decisions about its representation directly inform the design of information infrastructure, as Bowker and Star (2000) (among others) have shown. This infrastructure rapidly creates significant inertia and resistance to change: modifying data structures and contents directly impacts many of the purposes data are put to (which data actors become heavily invested in), skewing data representations heavily towards what has come before. How these implementations – and their reification in infrastructure – affect people's perceptions of one another and themselves is an intriguing and urgent question among a narrative of universal datafication. Recent work on activity tracking apps has shown that datafication of activity can come to take priority over personal experience – what you feel is what the data show, rather than vice versa (Littlejohns et al, 2019). As we further develop methods for analysing data configuration, examining the interactions between data, infrastructure, and (inter)personal perception will be rich soil to till.

As other sections of this chapter illustrate, decisions in the data funnel must be viewed in terms of the broader contexts in which those decisions are taken, and the underlying structures and perspectives that populate the players and the scene of the data collection stage. But it is tracing these decisions and reconstructing the erased historicity of data that enable us to understand how data contexts are materialized in data configurations, and to build new strategies for contesting and reimagining the configurations of our data ghosts.

Power dynamics in self-tracking data practices in the context of everyday life – Xiufeng Jia

Xiufeng identifies as a Chinese interdisciplinary researcher. She draws upon her background in AI, algorithms, and data in society, digital self-tracking, and digital health. Her current work focuses on responsible AI governance, design, and the everyday use of AI.

Data are associated with power dynamics (boyd and Crawford, 2012; Crawford et al, 2014; Van Dijk, 2014). These dynamics can involve political power, such as when data-driven technologies are used for governmental surveillance, or commercial power where data generate profits for corporations. Nonetheless, the power of data-generating individuals should not be ignored. Self-trackers, for example, create data but frequently gain relatively minimal benefits from their value (Sharon and Zandbergen, 2017). Their unique subjective experiences and narratives are crucial for understanding the value and meaning of data from their perspectives.

Ordinary individuals are configured as data subjects in self-tracking data practices, where their bodies and everyday lives are quantified into various types of data. These digital self-tracking technologies empower individuals to voluntarily collect, collate, analyse, and store their personal data. Through the quantified data, individuals strive to enhance self-knowledge and self-awareness regarding patterns in their physical exercise, food consumption, calories burned, menstruation cycles and fertility, mood, sleep, work productivity, daily expenses and other aspects of their bodies and lives in everyday life. They often rely on their personal data to understand and learn about their health conditions and bodily changes, managing their well-being either to maintain their current lifestyle and bodily situations or to improve upon them. In this sense, personal self-tracking data play a crucial role, giving people insights about how their bodies feel, what activities they engage with in their daily lives, and what decisions they need to make in their ongoing interactions with the data.

However, the intertwining of data and individuals in real-world scenarios is complex. Not all self-trackers strictly follow what their data suggest about themselves. Rather, individuals have capabilities to critically reflect on and analyse both themselves and their data in relation to their personal lived experiences and bodies. I argue that the agency of both individual and data is situated within the power dynamics of self-tracking data practices. These dynamics, often fluid and subtly present, are deeply embedded in real-life contexts. As Mascheroni (2020) suggests, understanding the everyday life contexts where data are embedded allows us to explore the power dynamics of datafication and its societal implications. Therefore, individualization and contextualization can be valuable for gaining a deeper understanding of power distribution and the agency of both individuals and data.

In particular, analysing individuals' feelings towards data and themselves presents a compelling approach to account for power relationships between data, bodies, self, and individual decision-making processes. Here, the term 'feelings' refers to emotions (for example, happiness and anxiety), bodily sensations (such as pain), and sensory experiences (for example, interactions with digital sensors). Some scholars (for example, Kennedy and Hill, 2018; Lupton et al, 2018) emphasize that people learn and behave through feelings. They suggest that feelings play an important role in understanding oneself and meanings of data to individuals. By comparing their data with their feelings, people develop embodied self-awareness and self-knowledge of their bodies.

To contextualize self-tracking data practices within the context of everyday life means to figure out individuals' feelings alongside their daily interactions with their data. Some self-tracking scholars (for example, Pantzar and Ruckenstein, 2017; Weiner et al, 2020) have stressed the importance of such contextualization for making sense of digital self-tracking data and oneself. This context becomes significant when individual self-trackers engage with their data within specific cultural and social everyday environments (Costa Figueiredo et al, 2017; Pantzar and Ruckenstein, 2017; Pink and Fors, 2017; Sharon and Zandbergen, 2017). For instance, some self-trackers feel their bodies becoming heavier, associating this with their recent dietary habits (for example, excessive intake of carbohydrates and sweets) and lifestyle changes (such as, ceasing cardio exercises). This leads them to review their relevant data such as calorie intake, weight, and physical exercise to validate their assumptions. This example demonstrates how daily habits and bodily sensations empower individuals to intuitively learn about their bodies, simultaneously indicating the powerful role of data for validating their bodily sensations, assuming they trust their data.

The interplay of the human body, data, and feelings creates a complex configuration of agency. This process shapes the construction of the self, demonstrating how individuals communicate with data and mutually engage with each other within these configurations. The ways in which self-trackers make sense of meanings of data, and use them, can be better understood within the context of their everyday lives, where they are living with their data on a regular basis.

Contextualizing the construction of older adults as data subjects – Nicole Dalmer

Nicole identifies as a queer, White settler. She draws upon her background in both Library and Information Science and critical gerontology, focusing on the role of connections (both to people and to technologies) in later life.

Digital technologies (and their associated data) are central to the shaping and (re)imagining of ageing futures. Older adults' bodies, routines, practices and behaviours are increasingly scrutinized, measured, and tracked as a means to foster healthy and active lifestyles, thus ensuring more cost-effective management of care in later life (Dalmer et al, 2022).

In exploring the multiple mechanisms that catalyse the creation of data subjects, it is imperative to identify and include those broader trends that are impacting specific population groups. The datafication of later life, for example, can be linked to four broader trends that are both contextualizing and shaping the experiences of ageing both in and with data.

The first is the imperative to age successfully. Successful ageing has gained prominence in and continues to inform theoretical paradigms, health measurements, retirement lifestyles, policy agendas, and anti-ageing ideals (Katz, 2013; Katz and Calasanti, 2015), with ageing successfully determined by: the avoidance of disease and disability, the maintenance of cognitive and physical function, and social engagement (Rowe and Kahn, 1997). Technologies (including wearables) are seen as tools to support older adults' capacity to age successfully. Digital tools (such as step counters and brain games), their data, and their 'smart' capacities can be used by older adults to self-track their performance and progress (via data points) to become more self-knowledgeable about and responsible for their well-being (Marshall, 2018). Of note, critics of this concept illuminate how successful ageing emphasizes individual responsibility (and choice) for one's lifestyle and successes (or failures) therein, without acknowledging the dynamics of power, opportunities, and inequalities.

The second broader trend is ageing in place. Ageing in place is a significant social, economic, and political goal, and refers to an individual ageing in their own home for as long as possible. Not only does ageing in place align with a majority of older adults' preferences, but it is also promoted as saving the public purse strings as it avoids costly relocation to institutional facilities. To support older adults' ability to age in place, technologies placed around the home are seen as tools to support older adults' independence in their homes and communities while simultaneously reducing (optimistically) family members' and other care partners' care work (Berridge, 2016). Consideration is less often given to the changes such digital devices (and the data they collect and transmit) bring to the feelings of privacy, security, and intimacy that are typically associated with one's home (Berridge and Wetle, 2020). The data outputted from such devices can also impact relationships, with adult children having unfettered access to and knowledge about their parents' habits and comings and goings. And so, while older adults prefer to age in their own homes, the tensions that digital devices in the home (that surveil, monitor, and report to others) introduce cannot be ignored.

The third trend, exacerbated by COVID-19, is the growing concern regarding social isolation among older adults. Older adults' experiences of social isolation and loneliness are an increasingly important topic of conversation in many countries, in part due to the many negative impacts of social isolation on older adults' physical and mental health and well-being: increased risk of premature death, depression, falls, cardiovascular disease, and dementia (Courtin and Knapp, 2017; Sepúlveda-Loyola et al, 2020). As evidence of an interventionist logic (Peine and Neven, 2019), technologies and associated transmission of data are proposed as interventions ('solutions') to solve the 'problems' of lonely or isolated older adults. Older adults are increasingly encouraged to share their personal data (via social media sites, email, photo sharing, and video chat tools) in order to improve their feelings of social connection (Petersen et al, 2023), thus also ensuring their capacity to age successfully.

Taken together, the imperative to age successfully and age in place, combined with the fears of social isolation in later life are changing the ways that bodies, homes, communities, and other spaces of care are understood; reframed as domains to be monitored, tracked, and managed via data to ensure that older adults' well-being and health are supported. In other words, we can see how these trends mould older adults as subjects of systems of governance.

Curiously, while the objective to support older adults to age cost effectively and successfully in place relies on the gathering and tracking of data, older adults themselves are often excluded from data gathering practices. Accordingly, the fourth and final trend that both creates and shapes older adults as data subjects are the many ageist stereotypes that are held about later life. Fernández-Ardèvol and Grenier (2022) have recently proposed the concept of data ageism[4] to convey the production and reproduction of the disadvantaged status of older age as a result of decisions concerning how, when, and where to collect and deliver data. Older adults, in data collection practices, are often treated as residual categories, which not only mutes the heterogeneity inherent in older adult populations, but can reinforce the invisibility of some older adults (Fernández-Ardèvol and Grenier, 2022). Given that data shape 'how and what we see' (Sendyka, 2013: 104), such exclusions can ultimately contribute to or perpetuate ageist thinking and practices.

Who is a (data) subject for autonomous vehicles? – Mergime Ibrahimi

Mergime identifies as a woman whose background is in critical data science, focusing on mobility, data diversity and perceptions towards automated systems.

Techno-companies and institutions that use technology often portray innovative data-driven technologies in a positive light, emphasizing their objectivity and efficiency in managing processes in different domains. Our understandings and imaginaries about autonomous vehicles (AVs), for example, are derived mostly from car companies' advertisements and thus create imaginary assumptions in the public that AVs will shape the future of mobility (Martin, 2021), and will solve the world's mobility problems. However, the ways in which people are represented (or fail to be represented) in data create contemporary debates about responsible and ethical use of AI. This perspective invites us to critically examine who gets to be defined as a relevant data subject for AVs.

Considering that AVs are envisaged to operate autonomously in complex and diverse societal settings, the configuration of data subjects becomes a critical question: Who is defined as a data subject – the passenger(s) inside the vehicle or other individuals in the surrounding environment? And how are the other individuals categorized and classified? On one hand, the passengers sitting inside the vehicle are the direct users of the technology who are subject to the decisions made by the AV. The configuration of these data subjects revolves around making sure of their safety, comfort, and convenience during transportation. On the other hand, there is the surrounding environment – other individuals around AVs such as pedestrians, cyclists, delivery robots, and other vehicles who need to both share the space with AVs and feel safe around them. It is AV designers who determine and define what encompasses the 'relevant' surrounding environment of AVs based on the input data they choose to incorporate into the algorithms. The AV algorithms are designed to process vast amounts of data rooted in a historical trajectory shaped by societal values, norms, and past technological advancements. Historical inequalities and biases can be inadvertently perpetuated in the data used to train these algorithms (Liu, 2017; Lim and Taeihagh, 2019), influencing how AVs perceive and respond to their surroundings. The lack of representation and agency for certain groups in historical data can have cascading effects on the configuration of data subjects in AV systems. Studies have revealed that AVs often fail to 'see' wheelchair riders or individuals with characteristics not present in the training data (Whittaker et al, 2019), as data diversity is often ignored. Also, research has identified potential biases within machine vision in general and the specific pedestrian detection algorithms as it tends to perform less accurately on females and individuals with darker skin tones (Buolamwini and Gebru, 2018), mirroring a long-standing history of hostile architecture and urban planning that rarely considers the needs of people with mobility impairments or marginalized communities.

One of the ethical challenges arising from identifying potential victims of car accidents is the determination of priorities. AVs must make split-second decisions when there might be an error in the system in potentially dangerous situations, such as choosing between crashing with an object or swerving to avoid it, which puts pedestrians or cyclists at risk. These decisions are based on complex algorithms trained on large datasets, but the lack of transparency in the training process and the specific data used raises questions about fairness, accountability, and the potential for bias.

Should AVs prioritize the safety of the passengers, considering the idea that users are paying for protection? Or should AVs prioritize the safety of pedestrians or other road users, upholding the principle of minimizing overall harm? Or should AVs save only the people whose data dominate the training datasets and eventually push certain other users off the roads? This conundrum extends beyond mere algorithmic calculations; it delves into the realm of moral and ethical reasoning (Robinson et al, 2022) and how that is perpetuated in AV algorithms.

Critically, this section does not propose that the overall goal of future work should be to simply lower the barriers to collecting and filtering the data. Rather, this section highlights a swath of normative questions about who becomes a data subject for AVs and which context data are considered essential in configuring these subjects. It has been argued that even if we use all available data, we still can get biased results because society is biased (Caliskan et al, 2017). This is because data-driven technologies are trained with data that tend to carry on the social dynamics and iterate 'the patterns of marginalisation, inequality, and discrimination' that exist in our societies, and are thus represented in the data (Leslie, 2019: 4). However, biases in data are not only generated by the discriminations that exist in society but rather by the contexts they are drawn upon, and the assumptions made during algorithm development. The filtering and selection of data plays a crucial role in shaping the outcomes and configurations of data subjects within AVs and other data-driven technologies.

In short, only those who are represented in the training data sets and have been historically considered 'relevant' are more likely to become data subjects that AV algorithms prioritize and consider in decision-making processes. This perpetuates a cycle of underrepresentation and exclusion, and highlights the need for a critical approach that considers the concerns and expectations of a broader range of data subjects, who often have very little agency in processes of technological governance.

Concluding remarks

The history, purposes, and contexts of data formation, collection, and use play fundamental roles in shaping the power and impact of data and data technologies. As we illustrate in this chapter, power dynamics are inextricably embedded in everyday data practices and in the configuration of individuals as data subjects.

The (re)configuring of people as data subjects within datafied systems and our broader datafied society is a complex process reflecting layered, multi-party histories of data construction and capture, design decisions and assertions of power, and technical implementation together with individual perception. Teasing out the nuances of power dynamics, notions and exercise of agency, and historical and contextual factors that underpin datafication requires multidimensional, transdisciplinary approaches. This chapter illustrates how such boundary-spanning approaches are vital to elucidate questions of how quantification of people in data occurs, and what the implications are for individuals and society.

Context plays a vital role in defining and shaping data subjects. The power dynamics within datafied systems are entwined with diverse and heterogeneous sociocultural environments, and these environments and the people within them shape every stage of data definition, collection, materialization, and use. Measuring people and perception based solely on 'universal' standards inevitably overlooks important contextual nuances, limiting our understanding of the impact of datafication on different communities and disparities in their input to datafication processes. By examining the feelings and experiences of individuals in their specific contexts, we can gain a deeper understanding of the power relationships between data, bodies, the self, and individual decision-making.

In each section of this chapter, we have invited the reader to question who becomes datafied, who enacts datafication, how this process is realized, and how datafication interacts with broader historical debates about personhood and the self. Taking up the findings from this chapter in conversation with the concept of the data funnel as illustrated in Figure 1.1: Denis' introductory section on the history of data begins to illuminate how the ways in which we define data and determine relevance come to bear on the *conceptualization* of data subjects; both Xiufeng's exploration of self-tracking practices and Nicole's broader examination of the datafication processes of a specific population (older adults) reveal the important questions that must be asked and decisions that must be made in the *conceptualization* and *observation* processes of the creation of the data subject (and how these decisions are materialized in data *recording*); and finally, Mergime's section critically examines who become data subjects for AVs and how the inherent biases and exclusions of potential data subjects are perpetuated by historical representation and relevance in training data sets.

This chapter reflects the multiplicity of datafication, both in the multiple perspectives that inform the representation of a single individual in data and in the multiple methodologies needed to critically examine and question the configuration of data subjects. To effectively engage with the process and politics of datafication, and the interwoven systems that configure individual data subjects, it is crucial to consider the historical context, power dynamics, and ethical implications that shape the collection, interpretation, and representation of data. People's individual perspectives, concerns, and expectations shape the power dynamics between themselves as data subjects and the datafication solutions that aim to encapsulate them. Data experts, as the creators and gatekeepers of data systems and the algorithms to analyse them, shape the data landscape. And insights into the often-inscrutable processes of algorithmic manipulation of data and data-driven decision-making is vital to understand how algorithms materialize datafication and its impacts. Examining the interplay between data subjects and data experts, the algorithms they create and are affected by, and the contexts they emerge from, operate in, and shape, are at the heart of understanding how people are configured as data subjects and imagining alternative (re)configurations.

DISCUSSANT RESPONSES

Individual and socio-technical practices of configuring data subjects – Katrin Amelang

Katrin responds to this nicely curated compilation deconstructing data subjects from the perspective of a science and technology studies inspired cultural anthropologist from Germany researching human–technology relations. In particular, she brings to the dialogue her recent work on the datafication of health and the body.

In critical data studies (CDS), one of the perhaps well-worn but also well-won commonplaces is that data are never raw or objective, merely depicting a neutral image of the world, but a specific, both partial and powerful way of capturing, knowing, and narrating it. Hence, CDS research emphasizes that digital data (sets) – like other results of knowledge production – have to be situated in particular times and places as well as in specific (data) practices and material-semiotic arrangements (for example, Loukissas, 2019; Kitchin, 2021). Against this backdrop and based on their respective research fields, the four authors focus on how individuals are made knowable in and through data, and thus on datafication processes, in which people are translated into digital data and are shaped as subjects of these data. While my research on mobile apps for menstrual self-observation provides another example supporting the authors' analysis of configuration processes of data subjects, I want to highlight and follow up on the socio-technical co-production of individuals/data subjects and data technologies.

The authors pay particular attention to the manifold conceptual choices made in the generation and collection of data and how these affect the configuration of data subjects. In this respect, Denis starts off with figuring datafication with the data funnel as decision and reduction processes, in which an individual framed as a specific data subject is narrowed down for the purpose of (specific) representation. All of them show such decisions are neither neutral nor random and anything but innocuous in their implications. For example, Nicole elaborates how sociocultural ideas, norms, stereotypes, and imaginaries about later life not only define older age as a specific problem and shape data collection and technologies designed for it, but also how older adults are grouped and homogenized on this basis, regardless of their different circumstances of life. Further, Mergime illustrates the conceptual power and historical legacies in the inclusion and exclusion of certain groups of people that, in her case of AVs, prioritize who becomes or does not become a data subject, and determine what becomes training data and thus implemented in technical innovations.

Adding the case of period-tracking, it is interesting who is being addressed as a menstruating person, potential app user, and data subject as well as how they are being imagined to be. The tracking categories these apps offer structure menstrual-self-monitoring practices and define what can (and cannot) be counted, tracked, and become data. The way apps engage users through user interface design and push-up notifications reveals their socio-technical scripts, which say as much about computational attempts at calculating menstrual cycles as they do about sociocultural ideas about menstruation and menstruating people or the politics of gender, sexuality, and reproduction, not to mention the politics and value of (menstrual) data. While the examples of all authors make clear that the conceptual history of data is inscribed, reflected, and materialized in the design of data systems, I would like to emphasize this point so as not to lose sight of the at once semiotic/social and material/technical constellations and configurations in datafication processes. Or, in Suchman's (2007) sense, data, data technologies, and data subjects are in a relationship of mutual constitution (and thus change).

Individuals are assembled and configured as data subjects not only by design through power-laden conceptualizations and realizations of data generation but also by encountering and using these data applications. Conversely, data technologies evolve with their uses and users. Not all data subjects are active users of data systems and not all actors putting data to use are the subject of these data (see Nicole's example of older adults). In this regard, the authors hint at the power of cooperative actors in datafication processes or indicate that options for decision and action are often not clear-cut and easily assignable. Xiufeng, in particular, deals with the power dynamics that unfold between data and individuals in the realm of self-tracking practices. In line with Sharon and Zandbergen (2017), Pantzar and Ruckenstein (2017), and Weiner et al (2020), she argues for embedding data practices in everyday life, considering meanings and values of data that not only reproduce but also challenge hegemonic notions of objective data truths, as well as paying more attention to the reflexive interplay of body data and feelings.

Again, I would emphasize technology (in this case, the app) in this far from unambiguous and frictionless interplay where people reconcile their datafied and embodied selves. The sensemaking of individual body experiences via app unfolds in a field of tension between normative pre-configurations (default settings) and moments of self-empowerment. In my conversations with users of period-tracking apps, the contingencies and contradictions of apps as instruments supporting self-knowledge become clear, but also how they deal with the ambivalent effects, gains, and uncertainties of data technologies in wayward, pragmatic, and reflexive ways (Amelang, 2022). Agential possibilities of data subjects may be more obvious in the case of individual, data-generating practices of self-tracking. Yet even in cases where individuals do not fit into, or are inadequately represented in, data systems (see Denis's example of applying for government benefits for the disabled), practices of failing, fitting into, or attempting to outsmart systems can be understood as part of the (re)configuration of individuals as data subjects. Emphasizing users does not mean forgetting other actors and data uses or underestimating the power of corporate and state actors, but does remind of the frictions and ambivalences in datafication processes (Hepp et al, 2022).

All this being said, I appreciate the deconstructing and critical reflexive approach of the authors. By exploring the question of who becomes a data subject and how, they introduce readers to various moments of both the construction of data and the configuration of data subjects. They thus also provide a starting point for further empirical research and for thinking how configurations of data subjects could be otherwise. This should include the role of technology in configuring the relation of data (subjects) and 'fleshy selves' as well as the moving beyond notions of accurate representation. Last but not least (and bridging to Doris), one should not forget that datafication always entails the promise of calculation, which translates between and relates individuals with populations and configures data subjects accordingly.

The inseparability of the individual from the collective – Doris Allhutter

Doris believes in the impact of collective subjectivity on research, but never came up with something that made sense to her when asked to identify. She writes this response with an interest in the entanglement of individual and collective subjectivity.

The authors of this chapter trace how datafication and data practices affect individuals and call to examine the configuration of data subjects through the lens of power dynamics, agency, and historical context. While reading the contributions, I wondered why it seems important to distinguish between the datafication of populations and individuals when historically grown power relations necessarily pervade the macro-, meso-, and micro-levels of society. My response attempts to trace and compare the power dynamics that the authors tease out in their respective cases.

Asking who gets to be defined as a relevant data subject for AVs, Mergime Ibrahimi points out the historical neglect of the needs of people with impairments and marginalized communities in urban planning and mobility (see Wilson et al, 2019). Privileges of safety and convenience may first go to solvent buyers of AVs. Technological governance, Mergime suggests, needs to be more inclusive and extend unequally distributed agencies to underrepresented communities. However, power would still side with the industry's individualistic vision of the future of mobility as private transport, as Mergime hints to in her introduction of the case. A moral dilemma perspective that ignores public value and argues case-by-case tolerates the ableist, racist, and classist power imbalances at play, I suggest.

Nicole Dalmer describes how older adults are constructed as subjects of governance systems to ensure more cost-effective management of care. She remarks that power dynamics influence people's capabilities of self-responsibility in 'successful ageing in place' and implicitly indicates intersectional identity categories such as gender, class, race, and ability. A critical ambivalence that points to restrictions on older adults' agency shows in the contradictory demands imposed on them. On the one hand, social isolation is seen as a matter of health concern that ought to be self-managed by maintaining social connections via social media. On the other hand, prioritizing self-responsibility and autonomy over care promotes that successful, data-supported ageing in place is an isolated activity. This individualization seems justified by the idea of a 'collective advantage-through-technology' (see Öchsner, 2021).

Xiufeng Jia describes self-trackers as data-generating individuals who aim to enhance their self-knowledge about data patterns derived from their bodies and daily practices. She defines power and agency as distributed between individuals and data and states that analysing individuals' feelings towards data and themselves accounts for power relations between data and self. Thinking with queer-feminist theories of affect and emotion, political theorist Brigitte Bargetz (2019) shows how affects and emotions point to the way in which people are imbricated in power relations. Affects and emotions do not indicate a subjective state, and they do not stand outside the social. Inequality produces specifically feeling subjects. Since data practices accommodate some people, bodies, and everyday practices more than others, they affect subjects differentially (Allhutter, 2021). We can start by asking who is invited to gain self-knowledge about their bodies and their everyday. And how do ideologies of human difference frame what kind of self-tracking practices address whom?

Finally, Denis Newman-Griffis focuses on the conceptual history of data. Data collected from a person applying for disability benefits mirrors a medicalized view of disability. It hardly considers information informed by the context-focused social model representing the person's disability experience. Down the data funnel, agency is distributed between the state, the benefits agents, and the applicants to ask and give information that influences the decision outcome, to dismiss some as irrelevant or apply discretion. Denis highlights the data's representational power, which affects people's perceptions of one another and themselves, and thus emphasizes the agentive historicity of data configurations, which are black-boxed by the erasure of the contexts of their emergence. Power also shows in the way representational bias is intimately linked with the material conditions of possibility for inclusion and exclusion.

The concluding remarks of the chapter focus the reader's attention on 'the power relationships between data, bodies, the self, and individual decision-making' by 'examining the feelings and experiences of individuals in their specific contexts'. This is a compelling approach. My reading tries to emphasize the richness of the authors' mobilization of power-critical perspectives. While centring on the configuration of individual data subjects, they offer a variety of entrance points to unpack power dynamics that operate through multiple layers of society. The datafication of populations and the datafication of individuals are powerful because they are inseparable.

Notes

[1] Discussant

[2] Discussant

[3] Facilitator

[4] Related, in part, to digital ageism, or age biases present in technologies, such as AI, where older adults may be excluded from technology development and design considerations as a result of prejudices or stereotypes (Manor and Herscovici, 2021; Chu et al, 2022).

References

Allhutter, D. (2021) 'Memory Traces in Society–Technology Relations: How to produce cracks in infrastructural power'. In: R. Hamm (ed) *Reader Collective Memory-Work*. Sligo: BeltraBooks, pp 426–452.

Amelang, K. (2022) '(Not) Safe to Use: Insecurities in everyday data practices with period-tracking apps'. In: A. Hepp, J. Jarke, and L. Kramp (eds) *New Perspectives in Critical Data Studies: The Ambivalences of Data Power*. Basingstoke & London: Palgrave Macmillan, pp 297–321.

Armstrong, S. (2017) 'Data, data everywhere: the challenges of personalised medicine', *BMJ*, 359.

Bargetz, B. (2019) 'A Political Grammar of Feelings: Thinking the political through sensitivity and sentimentality'. In: T. Bedorf and S. Herrmann (eds) *Political Phenomenology*. New York: Routledge, pp 52–70.

Barth, J., de Boer, W. E., Busse, J. W., Hoving, J. L., Kedzia, S., Couban, R., et al (2017) 'Inter-rater agreement in evaluation of disability: systematic review of reproducibility studies', *BMJ*, 356.

Beraldo, D. and Milan, S. (2019) 'From data politics to the contentious politics of data', *Big Data & Society*, 6(2): 2053951719885967.

Berridge, C. (2016) 'Breathing room in monitored space: the impact of passive monitoring technology on privacy in independent living', *The Gerontologist*, 56(5): 807–816.

Berridge, C. and Wetle, T. F. (2020) 'Why older adults and their children disagree about in-home surveillance technology, sensors, and tracking', *The Gerontologist*, 60(5): 926–934.

Bowker, G. C. and Star, S. L. (2000) *Sorting Things Out: Classification and its Consequences.* Cambridge, MA: MIT Press.

boyd, D. and Crawford, K. (2012) 'Critical questions for big data: provocations for a cultural, technological, and scholarly phenomenon', *Information, Communication & Society*, 15(5): 662–679.

Braun, M. (2021) 'Represent me: please! Towards an ethics of digital twins in medicine', *Journal of Medical Ethics*, 47(6): 394–400.

Buolamwini, J. and Gebru, T. (2018) 'Gender shades: intersectional accuracy disparities in commercial gender classification'. *Proceedings of the 1st Conference on Fairness, Accountability and Transparency*, PMLR (Proceedings of Machine Learning Research), 81: 77–91.

Caliskan, A., Bryson, J. J., and Narayanan, A. (2017) 'Semantics derived automatically from language corpora contain human-like biases', *Science*, 356(6334): 183–186. doi: 10.1126/science.aal4230. PMID: 28408601.

Chu, C. H., Nyrup, R., Leslie, K., Shi, J., Bianchi, A., Lyn, A., et al (2022) 'Digital ageism: challenges and opportunities in artificial intelligence for older adults', *The Gerontologist*, 62(7): 947–955.

Costa Figueiredo, M., Caldeira, C., Reynolds, T. L., Victory, S., Zheng, K., and Chen, Y. (2017) 'Self-tracking for fertility care: collaborative support for a highly personalized problem'. *Proceedings of the ACM on Human–Computer Interaction*, 1(CSCW): 1–21.

Courtin, E. and Knapp, M. (2017) 'Social isolation, loneliness and health in old age: a scoping review', *Health & Social Care in the Community*, 25(3): 799–812.

Crawford, K., Gray, M. L., and Miltner, K. (2014) 'Critiquing big data: politics, ethics, epistemology'. Special section introduction. *International Journal of Communication*, 8: 1663–1672.

Dalmer, N., Ellison, K., Katz, S., and Marshall, B. (2022) 'Ageing, embodiment and datafication: dynamics of power in digital health and care technologies', *International Journal of Ageing and Later Life*, 15(2): 77–101.

Fernández-Ardèvol, M. and Grenier, L. (2022) 'Exploring data ageism: what good data can('t) tell us about the digital practices of older people?', *New Media & Society*, 14614448221127261.

Gallicchio, S., and Bye, B. (1981) Consistency of initial disability decisions among & within states. No. 13. US Department of Health and Human Services, Social Security Administration, Office of Policy, Office of Research and Statistics.

Halpern, J. and Hausman, J. A. (1986) 'Choice under uncertainty: a model of applications for the social security disability insurance program', *Journal of Public Economics*, 31(2): 131–161.

Harrison, J. E., Weber, S., Jakob, R. and Chute, C. G. (2021) 'ICD-11: An International Classification of Diseases for the Twenty-First Century', *BMC Medical Informatics and Decision Making*, 21(6): 206. doi: 10.1186/s12911-021-01534-6.

Hepp, A., Jarke, J. and Kramp L. (2022) 'New Perspectives in Critical Data Studies: The Ambivalences of Data Power – An Introduction'. In: A. Hepp, J. Jarke, and L. Kramp (eds) *New Perspectives in Critical Data Studies: The Ambivalences of Data Power*. Basingstoke & London: Palgrave Macmillan, pp 1–23.

Katz, S. (2013) 'Active and successful aging: lifestyle as a gerontological idea', *Recherches sociologiques et anthropologiques*, 44(44–1): 33–49.

Katz, S. and Calasanti, T. (2015) 'Critical perspectives on successful aging: does it "appeal more than it illuminates"?', *The Gerontologist*, 55(1): 26–33.

Kennedy, H. and Hill, R. L. (2018) 'The feeling of numbers: emotions in everyday engagements with data and their visualization', *Sociology*, 52(4): 830–848.

Kitchin, R. (2021) *Data Lives: How Data Are Made and Shape Our World*. Bristol: Bristol University Press.

LaValle, S., Lesser, E., Shockley, R., Hopkins, M. S., and Kruschwitz, N. (2010) 'Big data, analytics and the path from insights to value', MIT Sloan Management Review.

Leslie, D. (2019) 'Understanding artificial intelligence ethics and safety', arXiv preprint, arXiv:1906.05684.

Lim, H. S. M. and Taeihagh, A. (2019) 'Algorithmic decision-making in AVs: understanding ethical and technical concerns for smart cities', *Sustainability*, 11(20): 5791.

Littlejohns, R., Gouthro, M. B., and Dickinson, J. (2019) 'Runners' engagement and social support practices: exploring the uses and role of online activities', *Sport in Society*, 22(12): 2243–2260. doi: 10.1080/17430437.2019.1571486.

Liu, H. Y. (2017) 'Three types of structural discrimination introduced by autonomous vehicles', *University of California Davis Law Review Online*, 51: 149–180.

Loukissas, Y. A. (2019) *All Data Are Local: Thinking Critically in a Data-driven Society*. Cambridge, MA: MIT Press.

Lupton, D., Pink, S., Heyes LaBond, C., and Sumartojo, S. (2018) 'Digital traces in context: Personal data contexts, data sense, and self-tracking cycling', *International Journal of Communication*, 12: 647–666.

Mabbett, D. (2005) 'Some are more equal than others: definitions of disability in social policy and discrimination law in Europe', *Journal of Social Policy*, 34(2): 215–233. doi: 10.1017/S0047279404008554.

Manor, S. and Herscovici, A. (2021) 'Digital ageism: a new kind of discrimination', *Human Behavior and Emerging Technologies*, 3(5): 1084–1093.

Marshall, B. L. (2018) 'Our Fitbits, Our (Ageing) Selves: Wearables, Self-Tracking and Ageing Embodiment'. In: S. Katz (ed) *Ageing in Everyday Life*. Bristol: Policy Press, pp 197–214.

Martin, N. (2019) 'How Much Does Google Really Know About You? A Lot', Forbes, 11 March. Available at: www.forbes.com/sites/nicolemart in1/2019/03/11/how-much-does-google-really-know-about-you-a-lot/ ?sh=73d856bb7f5d (accessed 16 July 2023).

Martin, R. (2021) 'AV futures or futures with AVs? Bridging sociotechnical imaginaries and a multi-level perspective of autonomous vehicle visualizations in praxis', *Humanities and Social Sciences Communications*, 8(1): 1–15.

Mascheroni, G. (2020) 'Datafied childhoods: contextualising datafication in everyday life', *Current Sociology*, 68(6): 798–813.

McFall, L., Meyers, G., and Hoyweghen, I. V. (2020) 'The personalisation of insurance: data, behaviour and innovation', *Big Data & Society*, 7(2): 2053951720973707.

Mooney, S. J., Westreich, D. J., and El-Sayed, A. M. (2015) 'Commentary: epidemiology in the era of big data', *Epidemiology*, 26(3): 390–394.

Neff, G. and Nafus, D. (2016) *Self-tracking*. Cambridge, MA: MIT Press.

Newman-Griffis, D. R., Hurwitz, M. B., McKernan, G. P., Houtrow, A. J., and Dicianno, B. E. (2022) 'A roadmap to reduce information inequities in disability with digital health and natural language processing', *PLOS Digital Health*, 1(11).

Newman-Griffis, D. R., Rauchberg, J. S., Alharbi, R., Hickman, L., and Hochheiser, H. (2023) 'Definition drives design: disability models and mechanisms of bias in AI technologies', *First Monday*, 28.

Öchsner, S. (2021) Is all ok? AAL as infrastructure for aging well. Doctoral thesis at the University of Vienna, Department of Science and Technology Studies.

Osterweis, M., Kleinman, A., and Mechanic, D. (1987) 'Conflicts and Contradictions in the Disability Program'. In: M. Osterweis, A. Kleinman, and D. Mechanic (eds) *Pain and Disability: Clinical, Behavioral, and Public Policy Perspectives*. Washington, DC: National Academies Press (US), pp 66–86.

Pantzar, M. and Ruckenstein, M. (2017) 'Living the metrics: self-tracking and situated objectivity', *Digital Health*, 3. doi: 10.1177/2055207617712590.

Peine, A. and Neven, L. (2019) 'From intervention to co-constitution: new directions in theorizing about aging and technology', *The Gerontologist*, 59(1): 15–21.

Petersen, B., Khalili-Mahani, N., Murphy, C., Sawchuk, K., Phillips, N., Li, K. Z. H., and Hebblethwaite, S. (2023) 'The association between information and communication technologies, loneliness and social connectedness: a scoping review', *Frontiers in Psychology*, 14. doi: 10.3389/fpsyg.2023.1063146.

Pink, S. and Fors, V. (2017) 'Being in a mediated world: self-tracking and the mind–body–environment', *Cultural Geographies*, 24(3): 375–388.

Robinson, J., Smyth, J., Woodman, R., and Donzella, V. (2022) 'Ethical considerations and moral implications of autonomous vehicles and unavoidable collisions', *Theoretical Issues in Ergonomics Science*, 23(4): 435–452.

Rowe, J. W. and Kahn, R. L. (1997) 'Successful aging', *The Gerontologist*, 37(4): 433–440.

Sambasivan, N., Kapania, S., Highfill, H., Akrong, D., Paritosh, P., and Aroyo, L. M. (2021) '"Everyone wants to do the model work, not the data work": data cascades in high-stakes AI', *Proceedings of the 2021 CHI Conference on Human Factors in Computing Systems*, pp 1–15.

Sendyka, R. (2013) 'Scopic regimes and modernity: hypotyposis'. In: D. Koczanowicz and D. Schauffler (eds) *Discussing Modernity: A Dialogue with Martin Jay*. Amsterdam & New York: Brill, pp 103–114.

Sepúlveda-Loyola, W., Rodríguez-Sánchez, I., Pérez-Rodríguez, P., Ganz, F., Torralba, R., Oliveira, D. V., and Rodríguez-Mañas, L. (2020) 'Impact of social isolation due to COVID-19 on health in older people: mental and physical effects and recommendations', *The Journal of Nutrition, Health & Aging*, 24: 938–947.

Sharon, T. and Zandbergen, D. (2017) 'From data fetishism to quantifying selves: self-tracking practices and the other values of data', *New Media & Society*, 19(11): 1695–1709.

Suchman, L. (2007) *Human-machine Reconfigurations: Plans and Situated Actions*. Cambridge: Cambridge University Press.

Van Dijk, T. A. (2014) *Discourse and Knowledge: A Sociocognitive Approach*. Cambridge: Cambridge University Press.

Weiner, K., Will, C., Henwood, F., and Williams, R. (2020) 'Everyday curation? Attending to data, records and record keeping in the practices of self-monitoring', *Big Data & Society*, 7: 1–15.

Werder, K., Ramesh, B., and Zhang, R. (2022) 'Establishing data provenance for responsible artificial intelligence systems', ACM Transactions on Management Information Systems (TMIS), 13(2): 1–23.

Whittaker, M., Alper, M., Bennett, C. L., Hendren, S., Kaziunas, L., Mills, M., et al (2019) 'Disability, bias, and AI', *AI Now Institute*.

Wilson, B., Hoffman, J., and Morgenstern, J. (2019) 'Predictive inequity in object detection', arXiv preprint, arXiv:1902.11097.

2

Children as Data Subjects: Families, Schools, and Everyday Lives

Karen Louise Smith, Leslie Regan Shade, Lyndsay Grant,
Priya C. Kumar, Lorenzo Giuseppe Zaffaroni, Gaia Amadori,
Giovanna Mascheroni, Marie K. Heath,[1] Daniel G. Krutka,[2]
Luci Pangrazio,[3] Neil Selwyn,[4] and Juliane Jarke[5]

Introduction

Writing in 2017, Lupton and Williamson noted that 'little research thus far has sought to examine how children are the objects of a proliferating range of digitized surveillance practices that record details of their lives' (Lupton and Williamson, 2017: 780–781). There are now various responses to this scholarly gap concerning the datafied lives of children (Holloway, 2019; Barassi, 2020; Grimes, 2021; Mascheroni and Siibak, 2021; Pangrazio and Mavoa, 2023). This chapter continues the response to this important area relevant to data power by presenting the results of a collaborative writing endeavour that brings together three distinct strands of empirical research situated in three contexts of datafied childhood.

Across various strands of research involving 'sharenting' discourse in the United States, privacy practices in Italian families, and the anticipatory data practices in secondary schools in England (UK), the foundations of neoliberalism, perceptions of risk, and decision-making against the backdrop of power relations permeate the work. The datafication of childhood reverberates with many decision-makers. Parents, teachers, governments, and others often need to make a variety of decisions in their everyday lives: should they use an app or 'smart' connected product? Do they want to opt-in or out of data collection? What should they do with data that have been collected from children? Recognizing these varied contexts, our research addresses some of the many issues facing datafied children, roughly falling between the ages of zero to 18.

Building upon concepts like data assemblages (Kitchin, 2014; Selwyn, 2021) and data futures (Dourish and Gómez Cruz, 2018; Ruppert, 2018), we have conceptualized the threads of our work as a braid. Although the threads of the datafication of childhood could be potentially infinite, Figure 2.1 teases out the three empirical themes upon which this chapter is based: sharenting, connected daily lives, and educational data analytics.

Following this introduction, our chapter begins with the first thread of the braid, sharenting. Kumar critically examines US media coverage of sharenting, a portmanteau of the words 'share' and 'parenting', that refers to parents posting about their children on social media. She uses the framework of governmentality to analyse nearly 250 articles, and she identifies potentiality risk as a neoliberal manifestation of data power. Potentiality risk encapsulates fears that data-driven processes will preclude children from maximizing their individuality and thwart any efforts to become their 'best selves'.

In the second thread, Zaffaroni, Mascheroni, and Amadori share insights from interviews with 20 Italian families with children aged 0–8. Their work identified multiple forms of connectivity (for example, smart phones, connected devices, educational technologies, and so on) through which children's personal information may be shared with other individuals, schools, and corporations. By examining distinct privacy contexts – interpersonal, commercial, and institutional – the study reveals the diverse challenges related to privacy that children and parents face in everyday life. Results show that parents hold different privacy practices and perceptions, ranging from privacy-protective strategies to surveillance realism, which reveal

Figure 2.1: Collaborative writing braid of datafied childhood

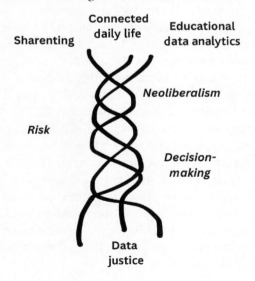

the contradictions, tensions, and anxieties of datafied childhood in family life where both risks and opportunities exist. These privacy practices and perceptions are grounded in a neoliberal context of individual responsibility, wherein parents are burdened with challenging and at times insurmountable responsibility for their children's data.

In the third strand of research, Grant shares ethnographic work on the anticipatory practices of data analytics that are found in English secondary schools in the United Kingdom. The work recognizes that the use of predictive data analytics to anticipate pupils' educational futures has intensified over the last decade, with emerging data-fuelled practices that identify certain pupils as 'risk subjects' to be managed and open to intervention. Neoliberal discourses concerning self-improvement are animated in real life when pupils are encouraged to position themselves on a trajectory towards a narrowly defined and predetermined, 'successful' future, continually prioritizing anticipated futures over present demands.

After exploring these three empirical threads, we move towards the idea of data justice as a potential resistance and response to the datafication of childhood. The data justice section contributed by Smith and Shade provides an opportunity to reflect upon some of the possibilities for intervention and action in response to intensive data collection and processing throughout childhood and beyond. This chapter argues that neoliberal notions of risk and decision-making permeate a variety of contexts of datafied childhood. We posit that while exploration of the datafication of childhood is still nascent, making visible commonalities across datafied childhood contexts provides a robust contribution towards striving for greater data justice.

The figure of the child: how fears about children's digital identities reinforce neoliberal individuality – Priya C. Kumar

Since the Internet entered the domestic sphere, academics, policy makers, and journalists have worried about the risks children face when using digital technologies (Livingstone, 2003; Banet-Weiser, 2004). Decades later, the proliferation of large-scale, data-driven, predictive modelling technologies throughout society raises considerable concern about how datafication can constrain children's lives (Lupton and Williamson, 2017; Barassi, 2020). These concerns are well founded, as the use of such technologies in schools and government agencies poses ethical questions and perpetuates inequities (Jones et al, 2018; Eubanks, 2019; Laird et al, 2022). But how does fear of datafication *itself* work to constrain children's lives? In other words, how do concerns about datafication coalesce into expectations about how children and caregivers should protect themselves and others from datafication's consequences?

I have approached this question through studying a practice that has attracted significant media and scholarly attention in recent years: parents posting pictures of children on social media. Colloquially called 'sharenting', a portmanteau of the words 'share' and 'parenting', the practice has been labelled as problematic and risky by journalists and researchers alike (for example, Kamenetz, 2019; Kopecky et al, 2020). At the same time, the practice is common: 82 per cent of American parents who use social media say they post about their children (Auxier et al, 2020). It also has historical precedent: families have exchanged photographs of children for nearly a century (Chalfen, 1987). I sought to trace what logic justifies concerns about this practice.

By analysing 243 US news articles on the topic using the framework of governmentality (Rose et al, 2006), which is a means of tracing how power works through expertise to regulate the lives of individuals, I found that sharenting discourse harnesses the concept of a digital identity to portray children's presence online as a form of risk (Kumar, 2021). This discourse treats the Internet as a distinct space of existence and equates posting about children with creating their digital identity. It references the consequences of this digital identity, expressing concern that data generated about a child now could affect them negatively in the future. Sharenting discourse states that children could experience physical, psychological, or emotional harm if someone kidnaps or bullies them based on information gleaned from their digital identity. This discourse also suggests that digital identities could foreclose children's future educational or employment opportunities, if college admissions officers or human resources staff negatively judge children based on these identities. This train of thought positions the digital environment not only as a separate sphere, but also one that threatens children. Yet temporality is key here, as sharenting discourse is primarily concerned with effects on the future child (or their adult self). In other words, the underlying concern of sharenting is that parental posting might thwart a child from reaching or achieving their potential.

Thus, I propose that sharenting discourse manifests a distinct form of risk, which I call *potentiality risk*. This risk encapsulates fears that data-driven processes preclude people from becoming their 'best selves'. It responds to anxiety that data-driven decision-making challenges people's agency, eroding their ability to chart their own life path. The neoliberal ideology that drives much of contemporary society impels people to exercise 'choice, autonomy, self-responsibility, and to maximize one's life as a kind of enterprise' (Rose et al, 2006: 91). Potentiality risk is the neoliberal recognition that data power threatens the project of maximizing one's individuality.

Sharenting discourse responds to this risk by holding parents responsible for mitigating the concerns of having a digital identity. It encourages parents to discipline their own behaviour, for instance, by 'thinking before you post', and to take steps to limit the visibility of children's information online. This discourse implies that parents who fail to take such preventative measures are at fault if (or when) their child experiences negative consequences linked to their digital identities. However, such logic ignores the fact that parents themselves have little control over the way children's information gets used. Indeed, even though corporate practice and legal regulation play a much more consequential role in governing data flows, sharenting discourse positions the commonplace activity of posting on social media as an individual concern with equally individual response.

Thus, I contend that 'sharenting' is not simply a word that describes what parents do. Rather, it is a discursive force that harnesses the concept of risk to distil social concerns of data power into an individual responsibility for preserving neoliberal selfhood. Concerns about sharenting come from a well-meaning place of wanting to protect children and respect their rights. However, efforts to resist data power must go beyond a focus on the actions of individuals and attend to the broader socio-technical conditions that give rise to datafication.

Privacy practices and perceptions in everyday family life – Lorenzo Giuseppe Zaffaroni, Giovanna Mascheroni, and Gaia Amadori

Italian families increasingly rely on Internet-connected devices and toys, which collect valuable insights into various aspects of children's lives – including play, communication, and education – turning them into datafied subjects (Barassi, 2020; Mascheroni and Siibak, 2021). However, our understanding of how the commercial and predictive use of children's personal data affects their social and educational outcomes is still limited (Mascheroni and Siibak, 2021). Concerns about privacy harms and security risks associated with children's data collection and use are widespread. While discussions often portray children and parents as unaware and powerless, further research should investigate how they actively shape their privacy practices and beliefs in response to datafication norms (Mascheroni, 2020).

Existing research on privacy in family life uncovered the tensions arising from various practices, including caring dataveillance (Leaver, 2017), sharenting (Blum-Ross and Livingstone, 2017), and the management of tracking technologies (Sukk and Siibak, 2021). Given these studies, it is now crucial to broaden our knowledge regarding the interconnected and context-dependent nature of privacy management within households. To this aim, we propose to frame privacy practices and perceptions as socially situated and emergent meaning-making processes rooted in everyday life. Building on Stoilova et al's (2019) framework, our study explores various data and privacy negotiations in daily interactions across three relevant contexts: interpersonal, institutional, and commercial. Understanding these distinct privacy contexts helps us grasp the varied privacy-related challenges experienced by children and parents, as different online environments shape specific expectations, values, and norms. We conducted in-depth interviews as part of a three-year longitudinal mixed-methods research involving 20 families with children aged 0–8 in the Milan metropolitan area, recruited through theoretical sampling. The participating families vary in terms of family composition, socio-economic status, cultural background, and religious participation. The interviews were coded and analysed using the Constructivist Grounded Theory methodology (Charmaz, 2014).

Our findings indicate that interactions among family members, family culture, and the media available at home significantly shape privacy practices and beliefs. Parents' digital privacy practices are shaped by a trade-off between adopting privacy-protective strategies against corporations or potential predators, and enabling children's access to technologies seen as crucial for their development and social connections. Privacy beliefs and practices in the context of the datafied home emerge within a framework of individual responsibility, whereby increasing neoliberal expectations are placed on parents as responsible mediators. Intimate surveillance (Leaver, 2017) is presented as essential to the social expectations and common norms around parenthood to which new parents are taught to conform. Furthermore, our findings reveal the central role of parental mediation, which evolves dynamically and informs privacy management depending on family history and culture, everyday dynamics, and relationships with other social groups and across different privacy contexts.

The context of interpersonal privacy is where parents address most of their privacy concerns and practices. Parents engage in discussions on interpersonal privacy by giving priority to the value of individual control, yet they hold various positions concerning whether or not to engage in privacy-sensitive practices (for example, sharenting). Indeed, interpersonal privacy is where parents exert more mediation over their children's media practices. Parents adopt a combination of enabling and restrictive mediation (Livingstone et al, 2017) in order to increase the safety of children's media experiences. For example, 'guest' or 'Kids' YouTube accounts are sometimes implemented as a way to make children's engagement with online media safer. While children develop 'smart' workarounds to communicate with friends or share content over (or created within) apps that are age-restricted (for example, TikTok), parents reveal their primary attention is towards interpersonal privacy against unknown others, and that they share a sense of unfamiliarity and disengagement with the 'inner' workings of various applications – that is, how they produce data traces for commercial purposes.

In the commercial context, parents often tend to adopt a 'surveillance realism' perspective (Hintz et al, 2018), thus normalizing corporate surveillance and profiling based on the premise they have 'having nothing to hide'. Parents employ both laissez-faire and restrictive strategies by safeguarding sensitive information against commercial risks, while downplaying the political and social consequences of datafication. Families' daily practices reveal the inconsistencies and uncertainties inherent in navigating surveillance capitalism. Often, paradoxical privacy beliefs and attitudes shape everyday decision-making. For example, some parents frame excessive smartphone use as a psychological issue by subscribing to social pressures and screen time discourses, while they simultaneously accept targeted advertisements and algorithmic decision-making as they ignore or dismiss the resulting biases or power imbalances.

Furthermore, within the commercial context, parents perceive less control over their data and privacy settings compared to the interpersonal context. This is partly due to the prevalent acceptance of surveillance capitalism and the belief that companies hold benign intentions. Parents often feel powerless in controlling commercial privacy settings, in contrast to the interpersonal context where they attribute more agency to themselves. However, parents may implement strategies to minimize children's exposure to profiling and surveillance, often adapting privacy approaches developed in the interpersonal context – thus resulting in various, sometimes inadequate, outcomes.

Regarding the institutional context, our findings indicate that parents generally have limited awareness of its implications. Privacy considerations are often absent when parents contemplate their children's digital interactions within an institutional datafied setting, such as the school. For instance, a 42-year-old mother approved the use of a school app called Kindertap primarily based on her trust in the school and in other parents, assuming that all of them had given their consent. Consequently, discussions regarding the data-driven processes of uploading, storing, and transmitting data through the app were not explicitly addressed. Seemingly, parental trust in the school overshadowed concerns related to institutional and commercial aspects, which were only superficially acknowledged.

In conclusion, our data highlight the situated and interconnected nature of privacy perceptions and practices, influenced by family cultures and media practices. Analysing different privacy contexts – interpersonal, commercial, and institutional – has provided valuable insights into the varying privacy implications and potential clashes that arise within these contexts as media practices unfold. By focusing on privacy in everyday life, we aim to shed light on new research directions that uncover the contradictions, tensions, and anxieties arising from the datafication of childhood and family life.

Pupils' data narratives: anticipating future risks – Lyndsay Grant

Schooling and education are sites in which relationships between the present and the future are brought to the fore, with education commonly positioned as an institution for preparing young people to successfully inhabit, or sometimes to create, imagined futures (Facer, 2021). The educational aims, practices, and governance regimes embedded in our schools are shaped by the challenges and opportunities we anticipate young people will face in the future. Education cast in this way, primarily as preparation for an anticipated future, is an exercise of power – those who claim knowledge over the future exercise authority over what we should do in the present to prepare for it (Facer, 2021).

The use of predictive data analytics to anticipate educational futures has intensified over the last decade, with large volumes of dynamically updated, automatically processed quantitative data used to support claims of more objective and accurate knowledge about future outcomes and scenarios (Mackenzie, 2015; Smithers, 2020, 2022; Jarke and Macgilchrist, 2021; Lunde, 2022). Such anticipatory data futures are increasingly powerful actors in the governance of education (Amsler and Facer, 2017; Webb et al, 2019). For pupils, predictive data analytics extrapolate from past performance data to anticipate and quantify future outcomes for both individuals and larger cohorts and demographics. In this way, data analytics constructs narratives connecting pupils' pasts and futures, building trajectories towards anticipated futures.

Exploring how education data are used to anticipate pupils' futures, I draw here on an ethnographic study of data practices in Ridgeway School, a secondary school (11–16 years old) in England (UK).[6] I conducted this research over the course of a full school year, following the generation of pupil performance data, its flows through and beyond the school, and its processing, representation, and implementation (Grant, 2022). Here, I argue that the educational data practices at work constructed narratives about who pupils were and where they were going, and were used to rationalize decisions and interventions applied to them in their present moment.

Understanding the high-stakes accountability and governance regimes at work in England provides an important context for the data practices working within this school. Schools' quality and performance are regularly inspected based, in large part, on pupils' performance data. The stakes for schools missing accountability targets are high: more frequent inspection, replacing their senior leadership, and joining a wider chain of sponsored academy schools. Headline school accountability targets are expressed in terms of the percentage of pupils each year achieving specified performance measures. But within schools, these cohort data measures are translated into individual targets for every pupil, with frequent data monitoring used to track their progress, anticipate and intervene in their future performance.

Ridgeway School contained a 'data office' dedicated to monitoring, analysing, intervening, and improving school and pupil performance data. Their key pupil performance monitoring and intervention system was represented in a 'data wall' displaying information on those pupils at risk of missing targets, comprising postcard-sized printouts containing information on each pupil including current and predicted performance data. Pupils were sorted and ranked in order of priority for being assigned to 'intervention' booster classes, with the aim of altering their predicted futures to better meet school targets.

This data wall, therefore, visualized and constituted pupils as 'risk objects' requiring active management through processes of rationing and prioritizing their assignment to intervention booster classes (Hardy, 2015; Wilkins and Gobby, 2022). Pupils positioned as risks to school accountability targets in this way were also rendered as opportunities for individual and school improvement. Through algorithmically anticipating the risks to accountability targets and rationalizing which pupils to assign to interventions, the school ultimately hoped to improve their future performance measures.

Pupils themselves were continuously made aware of their current position on the trajectory towards this narrowly specified future. For example, in every lesson, pupils selected a sticker detailing the learning objectives corresponding to their personal target level, and evaluated their work against this objective at the end of the lesson. So ingrained was this understanding of learning as 'working through the levels' (Livingstone and Sefton-Green, 2016) that when one teacher gave purely qualitative feedback on pupils' work in an attempt to prevent pupils, in his words, 'thinking of themselves as a number', pupils expressed deep dissatisfaction. Pupils were thus habituated to understand their own identity and learning primarily as progress through quantified levels towards a narrowly defined and predetermined future.

The progress targets upon which schools and pupils were evaluated were, at the time of fieldwork, calculated as three National Curriculum levels between the end of primary school (4–11 years old) and the end of secondary school (11–16 years old). A simple, linear trajectory was plotted between these points to determine pupils' 'expected' progress. In this way, future performance was assumed to be both predictable and calculable, despite educational research suggesting that most pupils do not, in fact, progress in such a constant, linear, and predictable fashion (Education Datalab, 2015). When pupils' data monitoring suggested they were diverging from this anticipated trajectory, they became eligible for interventions, withdrawing from arts and sports subjects to take booster classes in English and maths to catch up to where they 'should' be. The heavy reliance of schools on linear progress measures and booster classes has been described as fuelling an 'intervention culture', a form of dividing children in which some children are removed from the full range of lessons and school activities in order to 'plug gaps' in school data (Llewellyn, 2016; Bradbury et al, 2021). Here, these processes were facilitated, rationalized, and materialized through the anticipatory data practices at work in the pupil data wall.

By encouraging pupils to anticipate potential divergence from that path, pupils were asked to address themselves to a future that would always remain just out of reach. There would always be another level, a next future target to aim for; their achievement in the present moment would never be enough in and of itself, but only a stepping stone towards a future data point. In this way, the predictive data practices at work in this and many other schools, prioritized aspirational data futures over and above pupils' achievement, engagement, or pleasures in the present moment.

The narratives constructed through pupils' data trajectories constitute a form of risk rationality, producing pupils as risk objects, and rendering them amenable to be managed and intervened through data practices of triaging and prioritizing their assignment to intervention booster classes (Hardy, 2015; Wilkins and Gobby, 2022). Indeed, the rise of intervention cultures is part of the response to securing these anticipated risks. In these anticipatory data practices, children's futures are rendered as both known and knowable and as more valuable than their present, limiting opportunities for more meaningful engagement with their achievement, engagement, interests, and concerns in the present moment.

Data justice as resistance – Karen Louise Smith and Leslie Regan Shade

Various examples discussed thus far in this chapter – sharenting (Kumar), surveillance through everyday digital connectivity and smart devices in the home (Zaffaroni, Mascheroni, and Amadori), and intensifying uses of technology and data-driven metrics in the education system (Grant) – lead us towards the question of whether we can resist the datafication of childhood, which appears to be built upon a foundation of neoliberal ideals, which are both socio-technical. At this stage, we pivot to consider the possibilities that *data justice* offers as a form of resistance to configuring young people as data subjects.

Data justice is a concept that broadens data ethics issues and situates them within a social justice framework. As Lina Dencik, Arne Hintz, Joanna Redden, and Emiliano Treré argued, foregrounding social justice concerns can account for 'ongoing historical struggles against inequality, oppression and domination' (Dencik et al, 2019: 876), and provide a critical perspective on systemic and structural conditions that perpetuate inequalities and injustices. Data justice is sometimes approached from an intersectional feminist angle (for example, Taylor, 2017), which is significant as we consider the experiences of young people, because age is an attribute whose effects and experiences intersect with other identity characteristics.

By virtue of age, young people can encounter data injustices within groups, through platforms, or by organizations where they lack power. In many family contexts, it is parents who control the ebb and flow of posts through their sharenting, and they sometimes also select and configure the smart devices which collect and process their children's data. On social media platforms, young people are subject to the terms and conditions established by the legal teams for big tech corporations and, often, their content can be archived in near perpetuity. Similarly, many virtual learning and course management systems are controlled by companies that can track, profile, and surveil the academic progress of school children and university students. Finally, moving to the macro networked society level, young people's behavioural data are frequently tracked across platforms and subjected to algorithmic processing, which can shape online environments (targeted advertising, mis/disinformation), and access to opportunities (automated processing of job or university applications), often in discriminatory ways.

Various models of data justice including those from Taylor (2017) and the Advancing Data Justice Research and Practice Group (ADJRP) provide insights that are relevant to data justice and young people. Taylor's (2017) model of data justice, is supported by three pillars:

- *visibility* (access to representation, informational privacy, data as a public good);
- *technological engagement* (technological choice and autonomy, freedom to control one's data and to share in the benefits of data); and
- *non-discrimination* (the power to identify, and thus prevent and challenge biased and discriminatory data).

The ADJRP framework for data justice notes Western dominance. A 2022 literature review produced for The Alan Turing Institute in collaboration with The Global Partnership on AI (Leslie et al, 2022), cites Western dominance and biases in much of the work in data justice as a 'critical deficit' (p 8), and calls for 'relocat[ion of] data justice in the decolonial context' (p 14). ADJRP thus calls for a revisioning and reorientation of data justice along a framework encompassing six pillars: *power, equity, access, identity, participation*, and *knowledge*.

Ideas such as technological engagement from Taylor (2017) and participation in a decolonial context from ADJRP bring us closer towards more robust possibilities of data justice for children. Data justice for children must be inclusive of a broad spectrum of human rights. Myopic focus on data-centric issues and rights, such as individualistic Western models of 'consent' prevalent in privacy rights, when children often have little to no control over their surroundings (including at home and in school) have proven insufficient. Indeed, such a focus is problematic for everyone in light of the complexities and lack of transparency of data practices of technology corporations and governments. And practices such as data erasure or de-indexing, which are associated with the right to be forgotten, do little to meaningfully right the fundamental power imbalances involved. Data justice for children could likely be best supported by nations or other contexts where dialogue is fostered to appropriately regulate the situations in which data can be collected from children for commercial purposes, and to guard against discrimination and other algorithmic harms that could impact children. Data justice for children centres on their human rights and strives towards a more just society through fair and equitable access to digital technologies and skills.

Conclusion

In this chapter, we braided three threads of datafied childhood: sharenting, connected everyday lives, and educational data analytics. Concepts such as the potentiality risks of sharenting (Kumar), privacy protection and surveillance realism in family life (Zaffaroni, Mascheroni, and Amadori), and anticipating pupils' futures through data (Grant) reveal avenues for configuring children as data subjects, highlighting the need for children's data justice (Smith and Shade).

As we collectively address concerns of datafication, we must also attend to the way our responses may actually exacerbate, rather than alleviate, the problem. In the case of sharenting discourse, fears about parents' use of social media may intensify the already high levels of pressure and judgement that parents, especially mothers, face from society. Thus, I (Kumar) encourage future work to expand beyond a focus on individual action and towards ways to address the broader socio-technical conditions that give rise to concerns of data power.

We (Zaffaroni, Mascheroni, and Amadori) strongly advocate for additional research to delve into the situated and interconnected nature of privacy practices in the domestic context, particularly regarding privacy concerns in institutional and commercial contexts, and the different forms of data that family members produce and are (un)aware of. To address this shortfall, we propose to continue our focus on the implications of datafication for children's rights in the everyday life experiences of children and parents to gain valuable insights into the social ramifications and power dynamics associated with the datafication of childhood.

Anticipatory data futures risk foreclosing more open-ended possibilities for children's futures. I (Grant) argue that future research could productively engage with children's data in more open-ended, plural, and generative ways, rather than pre-empting pupils' futures and limiting opportunities in the present.

Data justice remains a future ideal for children who live increasingly datafied lives (Smith and Shade). To respond to the identified risks and power asymmetries in datafied childhood, given the persistence of the neoliberal context, two policy avenues are needed. One, regulatory guardrails from governments across jurisdictions, remain one of the most promising directions to support data justice; self-regulation by corporations has proven insufficient in many instances. Two, future policy should respond to the diverse lived experiences of children and include their voices in decision-making.

Discussants – Neil Selwyn and Luci Pangrazio

This chapter covers an admirable range of issues, and provides a great set of starting points for thinking further about children and young people as data subjects. In this concluding section we want to briefly flag a few points that might be taken forward in subsequent work:

1. *Developing more differentiated understandings of 'datafication' within the 0–18 years age span*

   The different threads in this chapter certainly cover a large span of young people's ages and stages. In moving forward, therefore, we need to pay closer attention to how data power is being differently encountered and experienced by the various cohorts that comprise the '0–18 years' umbrella. For example, how do data privacy contexts for a 15 year old differ from those of an 18 year old, or between toddlers and 'tweens'? What are the changing roles of parents and carers in managing and protecting their children's data as they grow and develop? While the chapter begins to give a sense of these age-related shifts, further work can be done to better understand the different privacy contexts experienced across the life course, and how they are learnt about and experienced.

2. *Scrutinizing the data economy that has grown up around children and young people*

   This chapter understandably focuses on some of the most immediate contexts (home, school) and familiar actors (parents, caregivers, teachers) implicated in children and young people's experiences of datafication. However, many other actors and processes also powerfully shape not only how young people encounter and make sense of data, but also the outcomes that these data may have in the future. In particular, in order to fully grasp the stories presented in this chapter we need to develop better understandings of the data economy – the corporate and technical drivers of where these young people's data goes, what it is used for, by whom, and with what outcomes. That said, it would be naive to assume that greater transparency of these machinations would necessarily ensure better outcomes. Greater understanding of the data economy is only of use if young people have the time and space to interpret this, and then a range of options or alternatives exist to ensure that it is indeed an actual choice. In contemporary society, both of these conditions are increasingly hard to find.

3. *Unpacking what we might currently critique as 'neoliberal' about datafication*
   A recurring point in the chapter's threads is the way datafication serves 'neoliberal' ends. This is an increasingly common conclusion arising from critical accounts of digitalization and datafication. Yet there is a need to give this further thought in future work – developing a more specific and detailed account of what is being argued, and avoiding the deployment of 'neoliberal' simply as a 'sloppy synonym' for the broad denouncement of modern capitalism (Ferguson, 2009). More specifically, then, the sense emerges from this chapter that data are now closely implicated in the techniques and modes of rationalization, auditing, and accounting that are used by power. Of importance here, then, is to now ask more specific questions about the outcomes arising from these forms of datafied rationalization, auditing, and accounting, and to dig further into exactly how these might be explicitly disadvantaging specific groups of children and young people. It is also worth exploring the ways in which digital data are entangled with broader power relations and the politics of oppression. In other words what is especially distinct, different, or new about these data-driven outcomes as a manifestation of neoliberal rationalization?

4. *Data justice as resistance … or complete reimagination of what might be?*
   The chapter (notably the closing contribution from Smith and Shade) appropriately frames data justice in terms of resistance, encouraging us to anticipate ways in which data might become a site of struggle rather than a site of surrender. That said, there is perhaps scope to also be more radical – that is, to engage in a complete rethinking of what forms of datafication might be desirable in young people's lives. In this spirit, it is worth pursuing what Evelyn Ruppert (2018) terms 'counter-imaginaries of datafication' to counter currently prevalent corporate ideas of what is thinkable about data power, therefore developing alternatives to dominant understandings of what practices are preferable and which actors we might want to be performing them. This is no easy task and will involve thinking further about the ontological and epistemological changes that would be necessary in our social institutions and society at large. Nevertheless, it is well worth retaining hope that the dominant forms of datafication currently to be found in young people's lives are *not* a done deal. Instead, we need to take the accounts present in this chapter as a basis from which to anticipate better data futures to come.

Discussants – Marie K. Heath and Daniel G. Krutka

This chapter artfully weaves together differing perspectives, emerging scholarship, and rich contexts to take on a critical topic of data power: the datafication of childhood. Through their collective braiding, the authors confront the forces at play on the phenomenon of children as data subjects and objects. We bring our background as critical scholars of education and technology to this discussion, picking up threads from this collectively written chapter. We offer a technosceptical lens to think through the role of technology, specifically digital data, in children's lives in and out of schools.

Unfortunately, everyday citizens, and especially children, have too often been afterthoughts in setting the terms for the technological terrain. Silicon Valley capitalists and edtech entrepreneurs often introduce technological 'innovations' with profit motives and without public dialogue or deliberation. The tech industry consistently forces consumers to react to a neoliberal technological world where individuals are left to fend for themselves. Early in Shoshana Zuboff's 2019 *The Age of Surveillance Capitalism* she shares the story of the development of what would be later known as 'smart' home technologies by computer scientists at the Georgia Institute of Technology (Georgia Tech). The researchers operated under the assumption that the data generated by the devices would remain in a 'closed loop', and the data 'would belong exclusively to the people who live in the house' (p 6). Silicon Valley capitalists such as Google disagreed, and the data from our 'smart' devices fuel massive industries and data extraction. Alphabet, Google's parent company, has been repeatedly sued for ignoring laws that protect children from this data exploitation.

Neil Selwyn and Luci Pangrazio offer a helpful definition for *neoliberalism* that accounts for how 'data is now closely implicated in the techniques and modes of rationalization, auditing, and accounting that are used by power'. What might it look like for people to join together to challenge and rein-in exploitative business models that dataify children – and all of us – without consent or consensus? The authors in this chapter offer vivid portraits that help us think through the problem, and how it should be framed. As educators, we similarly seek to set the boundaries of the debate.

We noticed a theme across the vignettes of not only neoliberalism, but also individualism, hypervisibility, and surveillance intruding upon childhood. In each example, a combination of adults, institutions, and technology work to pull children out of the immediacy of childhood's present and into an imagined future of adult creation. Priya Kumar named this *potentiality risk*, an imagined narrowing of a child's opportunity to maximize themselves as individuals because of their parents' choices, made, we would note, in a landscape of surveillance capitalism. Lorenzo Giuseppe Zaffaroni, Giovanna Mascheroni, and Gaia Amadori compared the ways parents navigate surveilling their own children and the ways institutions and commercial entities further surveil their children. Finally, Lyndsay Grant interrogated the transformation of children into numbers as a method to predict their academic futures.

We wonder, does datafication of children and students align with the values that members of our society hold about childhood and 'education'? A common refrain in educational technology circles holds that it is up to people to use technologies for good or bad. However, in his 1992 book *Technopoly*, Neil Postman argued that humans, often without awareness and conviction, have increasingly yielded their morals to the logics of machines. We become like our tools, which tends to work out well for those who profit from them. In his 1954 book *The Technological Society*, Jacque Ellul offered the term *technique* to describe 'the totality of methods, rationally arrived at and having absolute efficiency (for a given stage of development) in every field of human activity'.

With the normalization of surveillance technologies that capture and produce digital data, are we raising children to accept the logics of machines? Let us take Lyndsay Grant's description of the data wall at Ridgeway School where 'pupils were sorted and ranked in order of priority for being assigned to "intervention" booster classes, with the aim of altering their predicted futures to better meet school targets'. Whose vision of 'education' does this represent? Are there any educators, parents, or children who dream of this type of education? Or, does it satisfy the *technique* of machines for data, predictability, prediction? What does a data wall teach children about what is important in their community? And, what is important about themselves? What imagined futures are constrained or eliminated by reducing children to data?

Our classrooms and homes can offer spaces to remake our worlds by empowering young people to interrogate technology and resist. Instead of each individual trying to read unreadable Terms of Service to determine whether the online world is safe, communities can, as the authors of this chapter agree, stand for the common good. This means rejecting datafication (Pangrazio and Selwyn, 2021), discriminatory design (Benjamin, 2019), algorithmic bias (Noble, 2018), surveillance capitalism (Zuboff, 2019), and all other forms of technique that thwart human freedom and social justice. Fortunately, young people are particularly adept at questioning why things are as they are. They often still dream dreams of a more just world.

In our work through the Civics of Technology project (www.civicsoftechnology.org) and in our scholarship, we seek to struggle alongside young people for more just futures. Children's play often rejects *technique*. As teacher educators, we strive to understand what we should teach young people about data (curriculum), and how we should do so (instruction). We are guided by the assumption that 'technologies are not neutral and neither are the societies into which they are introduced'. Recognizing the force of technologies such as digital data, we utilize technosceptical questions as starting points. Drawing from Postman's 1998 talk (Krutka et al, 2022),[7] modified for this topic, we might ask:

- What do youth give up for the benefits of digital data?
- Who is harmed and who benefits from digital data?
- What does digital data need?
- What are the unintended or unexpected changes caused by digital data?
- Why is it difficult to imagine our world without digital data?

Each of these questions removes us from the neoliberal vision of data and instead towards a data justice movement of co-conspirators and abolition. Karen Louise Smith and Leslie Regan Shade argue for a robust vision of data justice for children that can be taken up at the national level. We agree with the need for developing 'counter-imaginaries of datafication' (Ruppert, 2018). We also contend that educators can advance a data justice curriculum in small acts of resistance and radical revisioning in our classrooms, neighbourhoods, and homes. Maybe in taking our data back, we can replace data walls with the dreams of young people.

Notes

[1] Discussant
[2] Discussant
[3] Discussant
[4] Discussant
[5] Facilitator
[6] Names have been pseudonymized to protect anonymity
[7] www.civicsoftechnology.org/curriculum

References

Amsler, S. and Facer, K. (2017). Contesting anticipatory regimes in education: Exploring alternative educational orientations to the future. *Futures*, 94: 6–14. https://doi.org/10.1016/j.futures.2017.01.001

Auxier, B., Anderson, M., Perrin, A., and Turner, E. (2020). Parents' attitudes – and experiences – related to digital technology. Pew Research Center. www.pewresearch.org/internet/2020/07/28/parents-attitudes-and-experiences-related-to-digital-technology/

Banet-Weiser, S. (2004). Surfin' the Net: Children, Parental Obsolescence, and Citizenship. In M. Sturken, D. Thomas, and S. Ball-Rokeach (eds), *Technological Visions: The hopes and fears that shape new technologies* (pp 270–292). Temple University Press.

Barassi, V. (2020). *Child Data Citizen: How tech companies are profiling us from before birth*. MIT Press.

Benjamin, R. (2019). *Race after Technology: Abolitionist tools for the New Jim Code*. John Wiley & Sons.

Blum-Ross, A. and Livingstone, S. (2017). 'Sharenting,' parent blogging, and the boundaries of the digital self. *Popular Communication*, 15(2): 110–125.

Bradbury, A., Braun, A., and Quick, L. (2021). Intervention culture, grouping and triage: High-stakes tests and practices of division in English primary schools. *British Journal of Sociology of Education*, 42(2): 147–163. https://doi.org/10.1080/01425692.2021.1878873

Chalfen, R. (1987). *Snapshot Versions of Life*. Bowling Green State University Popular Press.

Charmaz, K. (2014). *Constructing Grounded Theory* (2nd edn). SAGE.

Dencik, L., Hintz, A., Redden, J., and Treré, E. (2019). Exploring data justice: Conceptions, applications and directions. *Information, Communication & Society*, 22(7): 873–881. https://doi.org/10.1080/13691 18X.2019.160626

Dourish, P. and Gómez Cruz, E. (2018). Datafication and data fiction: Narrating data and narrating with data. *Big Data & Society*, 5(2): 2053951718784083.

Education Datalab. (2015). Seven things you might not know about our schools (pp 1–30). Education Datalab. http://educationdatalab.org.uk/2015/03/seven-things-you-might-not-know-about-our-schools/

Ellul, J. (1954/2021). *The Technological Society*. Vintage.

Eubanks, V. (2019). *Automating Inequality: How high-tech tools profile, police, and punish the poor*. St. Martin's Press.

Facer, K. (2021). Futures in Education: Towards an ethical practice. Paper commissioned for the UNESCO Futures of Education report (pp 1–29). UNESCO Futures of Education Commission/ERF Working Paper for UNESCO. https://unesdoc.unesco.org/ark:/48223/pf0000375792.locale=en

Ferguson, J. (2009). The uses of neoliberalism. *Antipode*, 41(S1): 166–184.

Grant, L. (2022). Reconfiguring Education through Data: How Data Practices Reconfigure Teacher Professionalism and Curriculum. In A. Hepp, J. Jarke, and L. Kramp (eds), *New Perspectives in Critical Data Studies: The ambivalences of data power* (pp 217–239). Palgrave Macmillan. https://doi.org/10.1007/978-3-030-96180-0_10

Grimes, S. M. (2021). *Digital Playgrounds: The hidden politics of children's online play spaces, virtual worlds, and connected games*. Toronto: University of Toronto Press.

Hardy, I. (2015). Data, numbers and accountability: The complexity, nature and effects of data use in schools. *British Journal of Educational Studies*, 63(4): 467–486. https://doi.org/10.1080/00071005.2015.1066489

Hintz, A., Dencik, L., and Wahl-Jorgensen, K. (2018). *Digital Citizenship in a Datafied Society*. Polity.

Holloway, D. (2019). Surveillance capitalism and children's data: The internet of toys and things for children. *Media International Australia*, 170(1): 27–36.

Jarke, J. and Macgilchrist, F. (2021). Dashboard stories: How narratives told by predictive analytics reconfigure roles, risk and sociality in education. *Big Data and Society*, 8(1): 1–15. https://doi.org/10.1177/20539517211025561

Jones, K. M., McCoy, C., Crooks, R., and VanScoy, A. (2018). Contexts, critiques, and consequences: A discussion about educational data mining and learning analytics. *Proceedings of the Association for Information Science and Technology*, 55(1): 697–700. https://doi.org/10.1002/pra2.2018.1450 5501085

Kamenetz, A. (2019, June 5). The Problem With 'Sharenting.' *The New York Times*. www.nytimes.com/2019/06/05/opinion/children-internet-privacy. html

Kitchin, R. (2014). *The Data Revolution: Big data, open data, data infrastructures and their consequences*. SAGE.

Kopecky, K., Szotkowski, R., Aznar-Díaz, I., and Romero-Rodríguez, J.-M. (2020). The phenomenon of sharenting and its risks in the online environment: Experiences from Czech Republic and Spain. *Children and Youth Services Review*, 110: 104812. https://doi.org/10.1016/j.childyouth. 2020.104812

Krutka, D. G., Metzger, S. A., and Seitz, R. Z. (2022). 'Technology inevitably involves trade-offs': The framing of technology in social studies standards. *Theory & Research in Social Education*, 50(2): 226–254.

Kumar, P. (2021). From Oversharing to Sharenting: How Experts Govern Parents and Their Social Media Use (Doctoral dissertation, University of Maryland, College Park).

Laird, E., Grant-Chapman, H., Venzke, C., and Quay-de la Vallee, H. (2022). Hidden Harms: The Misleading Promise of Monitoring Students Online. Center for Democracy and Technology. https://cdt.org/wp-cont ent/uploads/2022/08/Hidden-Harms-The-Misleading-Promise-of-Mon itoring-Students-Online-Research-Report-Final-Accessible.pdf

Leaver, T. (2017). Intimate surveillance: Normalizing parental monitoring and mediation of infants online. *Social Media + Society*, 3(2). https://doi. org/10.1177/2056305117707192

Leslie, D., Katell, M., Aitken, M., Singh, J., Briggs, M., Powell, R., et al (2022). Advancing data justice research and practice: An integrated literature review. The Alan Turing Institute in collaboration with The Global Partnership on AI. https://arxiv.org/pdf/2204.03090.pdf

Livingstone, S. (2003). Children's use of the internet: Reflections on the emerging research agenda. *New Media & Society*, 5(2): 147–166. https:// doi.org/10.1177/1461444803005002001

Livingstone, S. and Sefton-Green, J. (2016). *The Class: living and learning in the digital age* [e-book]. New York University Press. http://opensquare. nyupress.org/books/9781479824243/

Livingstone, S., Ólafsson, K., Helsper, E. J., et al (2017). Maximizing opportunities and minimizing risks for children online: The role of digital skills in emerging strategies of parental mediation. *Journal of Communication*, 67(1): 82–105.

Llewellyn, A. (2016). Problematising the pursuit of progress in mathematics education. *Educational Studies in Mathematics*, 92(3): 299–314. https://doi.org/10.1007/s10649–015–9645–8

Lunde, M. I. (2022). Learning analytics as modes of anticipation: Enacting time in actor-networks. *Scandinavian Journal of Educational Research*, 68(2): 218–232. https://doi.org/10.1080/00313831.2022.2123851

Lupton, D. and Williamson, B. (2017). The datafied child: The dataveillance of children and implications for their rights. *New Media & Society*, 19(5): 780–794. https://doi.org/10.1177/1461444816686328

Mackenzie, A. (2015). The production of prediction: What does machine learning want? *European Journal of Cultural Studies*, 18(4–5): 429–445. https://doi.org/10.1177/1367549415577384

Mascheroni, G. (2020). Datafied childhoods: Contextualising datafication in everyday life. *Current Sociology*, 68(6): 798–813. https://doi.org/10.1177/0011392118807534

Mascheroni, G. and Siibak, A. (2021). *Datafied Childhoods*. Peter Lang.

Noble, S. U. (2018). *Algorithms of Oppression*. New York University Press.

Pangrazio, L. and Mavoa, J. (2023). Studying the datafication of Australian childhoods: Learning from a survey of digital technologies in homes with young children. *Media Information Australia*, 0(0). https://doi.org/10.1177/1329878X231162386

Pangrazio, L. and Selwyn, N. (2021). Towards a school-based 'critical data education'. *Pedagogy, Culture & Society*, 29(3): 431–448.

Postman, N. (1992). *Technopoly: The surrender of culture to technology*. Vintage.

Rose, N., O'Malley, P., and Valverde, M. (2006). Governmentality. *Annual Review of Law and Social Science*, 2(1): 83–104. https://doi.org/10.1146/annurev.lawsocsci.2.081805.105900

Ruppert, E. (2018, 14 June). Sociotechnical imaginaries of different data futures: An experiment in citizen data. 3e Van Doornlezing. www.eur.nl/sites/corporate/files/2018-06/3e%20van%20doornlezing%20evelyn%20ruppert.pdf

Selwyn, N. (2021). Key concepts: The 'data assemblage' [Website]. The Data Smart School. https://data-smart-schools.net/2021/05/21/key-concept-the-data-assemblage/

Smithers, L. (2020). Student success as preemption: Predictive constructions of futures-to-never-come. *Futures*: 124. https://doi.org/10.1016/j.futures.2020.102639

Smithers, L. (2022). Predictive analytics and the creation of the permanent present. *Learning, Media and Technology*: 1–13. https://doi.org/10.1080/17439884.2022.2036757

Stoilova, M., Nandagiri, R., and Livingstone, S. (2019). Children's understanding of personal data and privacy online: A systematic evidence mapping. *Information, Communication & Society*, 24(4): 557–575. https://doi.org/10.1080/1369118X.2019.1657164

Sukk, M. and Siibak, A. (2021). Caring dataveillance and the construction of 'good parenting': Estonian parents' and pre-teens' reflections on the use of tracking technologies. *Communications*, 46(3): 446–467. https://doi.org/10.1515/commun-2021–0045

Taylor, L. (2017, July–December). What is data justice? The case for connecting digital rights and freedoms globally. *Big Data & Society*, 1–14. https://doi.org/10.1177/2053951717736335

Webb, P. T., Sellar, S., and Gulson, K. N. (2019). Anticipating education: Governing habits, memories and policy-futures. *Learning, Media and Technology*, 0(0): 1–14. https://doi.org/10.1080/17439884.2020.1686015

Wilkins, A. and Gobby, B. (2022). Objects and subjects of risk: A governmentality approach to education governance. *Globalisation, Societies and Education*, 0(0): 1–14. https://doi.org/10.1080/14767724.2022.2114073

Zuboff, S. (2019). *The Age of Surveillance Capitalism: The fight for the future at the new frontier of power*. Hachette Book Group.

3

In/visibilities in Data Studies: Methods, Tools, and Interventions

Miriam Fahimi,[1] Petter Falk, Jonathan W. Y. Gray,[2] Juliane Jarke,[3]
Katharina Kinder-Kurlanda, Evan Light, Ellouise McGeachey,
Itzelle Medina Perea, Nikolaus Poechhacker, Lindsay Poirier,
Theo Röhle, Tamar Sharon, Marthe Stevens, Bernard van Gastel,
Quinn White, and Irina Zakharova

Introduction

Research in data studies is rarely a straightforward pursuit and can lead one down existential pathways of both angst and joy. While the idea of 'data' may present itself as something simple enough to parse and probe, 'data' – their production, use, analysis, and manipulation – are complex to grapple with. The bounds of data are ambiguous; the meaning of data is polysemous, and the social impacts of data are diverse. Just as renderings of data focus attention on some things while eclipsing others, fixing any one critical analytical lens on data can provide insight on certain socio-political issues while inevitably precluding others from view. Rendering data visible in all its fine details thus necessarily instigates, provokes, and challenges power. For example, data are not only held but also defined by gatekeepers (for example, Big Tech corporations, governments) which makes access to and openness of data political. Diverging perspectives on which data count and the interventions to make them count or make the gatekeepers accountable provide another example.

These power imbalances underlie our explorations into methods, (conceptual) tools and interventions within data studies. This chapter examines data in/visibilities through a multi-perspectival lens – differentially rendering data with each turn of the kaleidoscope. The kaleidoscope is a tool that allows for multiple perspectives, and seemingly multiplies the reality of what we see.

Without it, the multiplicity of perspectives would not be possible. Bringing to the fore these different perspectives, we show how acts of rendering data visible – whether as data publishers, activists, or data studies scholars – quickly lead us to interrogating power dynamics and power imbalances. The sections meditate on a number of questions: How do we chart the contours of visibility for something as ephemeral and capricious as data? How do we evaluate what we can see – its truth, its accessibility, and its social value? How do we make sense of when and for whom data visibility is productive?

The sorts of data we probe in this chapter are diverse – ranging from government datasets to corporate databases to administrative algorithms. Sections of the chapter also fix their analytical lenses on diverse sites in a data lifecycle and interrogate the roles of a diverse set of gatekeepers in shaping data's meaning and perceptibility. The chapter is organized in three parts, each interrogating the intersections between in/visibility and power. In Part 1, authors discuss the challenges of (academic) knowledge production about and with data and explore ways to capture data's relativity, fluidity, and instability as well as trace and visualize the movement of data. For example, frameworks for studying frictions in the (often invisible) infrastructures of algorithms render visible actors, technologies, and their diverging values in handling data. In Part 2, authors offer empirical case studies of data renderings – analysing the consequences of data visibility while also reflecting on the methodological opportunities and challenges of foregrounding the values and norms encoded into data. In the final part, contributors outline tool-based interventions for bringing alternative data framings and narratives into visibility.

PART 1: CONCEPTUAL APPROACHES TO DEALING WITH DATA

Methodologies and performativities of data studies – Irina Zakharova[4]

The emerging interdisciplinary field of data studies is concerned with the role of digital data, algorithms, and automation in society. While research on digital data benefits from the variety of disciplinary, theoretical, and methodological approaches to digital data, scholars of datafied societies also deal with particularly complex empirical questions. Typically, any kind of research in social sciences needs to accommodate and explain complexity of a given empirical field, be it public administration, health, or research on everyday lived experiences of individual people. When studying datafied societies, however, researchers face several conceptual and practical challenges reflecting the characteristics and contingencies of digital data.

For example, one challenge emerges from an understanding of digital data as relational products of negotiations between various actors – technology designers, policy makers, funders, and, sometimes, potential technology users or communities affected by technology design – rather than simple representations of social reality. Another conceptual challenge is in uncovering whose and which values and dispositions about the social reality certain data mean to represent. Furthermore, scholars and their study participants have to determine what data or algorithms are in a given research context. Digital data are not stable objects, the lifecycle of which can be followed through from their generation to their use and eventual decay, even the practitioners whose work we study struggle to provide clear definitions. In each step of their lives, data change forms, formats, ownership, and meanings in accordance with the individual, organizational, and political goals of the actors currently engaged with these data. Often the access to these data at different stages of their lives can be restricted through organizational and legal regulations, and by actors who own (access to) these data, while particularly public and nongovernmental organizations increasingly commit to opening their and others' data for further examination and use. To access and analyse such data and related technologies, researchers either rely on the resources and instruments provided by technological corporations or develop their own tools and methods.

The choice and selection of such instruments and methods is performative to what we know about the datafication of society. This performativity of research methods discussed by scholars of science and technology is crucial for the future paths of both academic and technology development (Law, 2004; Law et al, 2011; Ruppert et al, 2013). Usually, performativity is acknowledged as the potential of research activities to 'meddle' with the empirical reality, as methods textbooks and handbooks warn (early career) scholars about biases in research practice or advise on navigating research ethics and transparency. Instead, I argue here for taking performativity of research as a starting point in the methodological reflection on data studies. If we do not know the boundaries of our knowledge and do not reflect on how these boundaries come to be, it is quite difficult to imagine alternative paths for future development. Currently, various scholars engage with such a reflection given the pressing need for new, collective, and just futures for living with digital and automated technology (Dencik et al, 2022; Powell et al, 2022; Zakharova, 2022).

I argue that one way to reflect on the performativity of data studies is to combine methodological reflection with specific questions related to challenges in studying digital data. The performativity approach allows making visible which 'agential cuts' (Barad, 2007) scholars draw to conduct empirical studies. Drawing on John Law's methodological exploration of research practices in social sciences, there are four core interrelated elements of the research design (Law, 2004). These elements include, first, the researchers and how they position themselves within relevant academic fields and empirical sites of practice. Second, how the researched persons and objects in their historical development constitute the object of study. Third, particular research processes, instruments, and procedures as well as guiding research politics and organizational contexts, such as research funding, comprise the bandwidth of practical factors shaping research design. Finally, the empirical site of practice where a particular research project is conducted provides empirical context of the methodological reflections.

To incorporate the challenges of data studies in the framework of the methods assemblages, the performativity perspective is crucial. Instead of searching for a unified definition of what digital data and related algorithmic systems are or should be, a performative approach to data studies embraces the multiplicity and uncertainty of societal datafication processes. The individual, situated, contextualized knowledge of what digital data stand for in particular research situations is what draws each method assemblage together and provides researchers with sensitivities to render their approaches visible and understandable in the interdisciplinary field that is data studies. The goal of such reflection is to develop new vocabularies that loosen the long-standing binaries of technology/society, nature/culture, research/practice.

In/visibilities of algorithmic infrastructure: frictions as methodological lens for algorithmic ethnography – Miriam Fahimi, Nikolaus Poechhacker, and Katharina Kinder-Kurlanda

Studying algorithms in development poses a methodological challenge for ethnographic research (Kitchin, 2017). Ethnographers need to deal with different moments of in/visibility of algorithmic systems, which are opaque black boxes (Burrell, 2016; Christin, 2020), diffuse and heterogeneous (Seaver, 2017), and even mysterious (Ziewitz, 2016) things for the ethnographer that undertakes to study them. Determining where algorithms begin and end, as well as the mechanisms that lead to the stabilization of values and norms, is therefore often challenging to observe. In such highly mediated environments, the modes of observation have to adapt, and ethnographers require specific expertise to be able to follow and understand practices with and by computers.

One possible approach to ethnographically grasping opaque and 'boring things' (Star, 1999) as algorithms is by thinking of them as infrastructure. Infrastructures sustain social structures (Bhaskar, 2008), they are a representation of the world, and they can encode inequalities (Star, 1999). They are especially effective, and tend to become unquestioned once established (Katzenbach, 2021). However, in the moment of their breakdown not only does infrastructure become visible again, but so do social norms and assumptions conveyed therein. Yet, it remains unclear in which situations an algorithmic infrastructure 'fades away' (Star, 1999).

When we understand algorithmic infrastructure as relational (Star and Ruhleder, 1996), friction – the tiny moments of breakdown during continuous maintenance efforts – represent exciting moments for ethnographers. Drawing on our own positionality as ethnographic researchers and science and technology studies scholars, we experienced how social norms and values on gender in a credit scoring system became visible after paying specific ethnographic attention to their mediation by the infrastructure and multiple situations of friction: there, in a European credit agency, personal data were stored in a database, including people's name, address, date of birth, and a (binary) gender entry. While all personal data in this database were editable, such as name changes following marriage, address changes resulting from relocation, or corrections of typographical errors in date of birth, the gender entry could not be edited, even in situations where corrections were necessary. While these frictions had not been recognized as such by the computational practitioners for decades, this changed with public debate and the introduction by law of a third gender option. In our case, it created friction for our interlocutors, namely when a gender entry was entered incorrectly, or when a person legally changed their gender entry. As a result, the practitioners started several attempts at tinkering with the database, for example, by making the 'correct' gender entry visible through a commenting function that was added. This can be seen as an ongoing maintenance of the infrastructure with the aim to keep it up to date with changing normative (e)valuations (Graham and Thrift, 2007). And yet, these maintenance efforts did not translate over to the logic of the scoring algorithm. The algorithm could not process the comments, but only the original data entries, to describe the person's gender. Thus, the efforts to make misclassifications visible via tinkering and ex-post workarounds did not carry over to the actual computational model. This friction was eventually smoothed out by the introduction of a new IT system, which then allowed for changes of gender entries and the inclusion of a third gender option as a variable. We conceptualized this as a second more impactful moment of tinkering and maintenance: while the first moment was important to make the problem visible, the second one created (again) a seamless integration of the now changed normative account on gender.

Friction in practices related to algorithmic infrastructures creates both moments of crisis and tinkering. Subsequently, we consider posing the following questions as insightful for ethnographic research in data-dense environments:

- *Tiny moments of breakdown*: In which practices do algorithmic systems produce friction? Which (former invisible and taken-for-granted) norms, values, and works become visible?

- *Fixing algorithmic infrastructure*: How and by which relevant actors are such frictions addressed, solved, tinkered with, fixed? Which novel problems arise while and after 'fixing'?

To conclude, we highlight that ethnography is also a way to intervene in the field of computational predictions, as it sensitizes us to how norms and values fold and unfold in algorithmic systems. By raising awareness of the moments of friction we can also become active agents of change in the complex interplay of making issues and their solutions in/visible.

Challenges of tracing and visualizing data journeys – Itzelle Medina Perea

One approach that can be helpful to explore the circulation of data is 'data journeys' (Bates et al, 2016). This methodology offers a way to examine the movement of data across space and time, through different cultures and sites of practice from their initial generation through to reuse in diverse contexts.

Data journeys has proven to be useful in tracing data flows in different contexts (for example, Swist et al, 2019), and the insights gained through this exploration are useful to inform the design of accessible visualizations and descriptions which can contribute to efforts of making data flows more transparent (Bates et al, 2023). However, a number of challenges may arise when implementing this methodology. In this contribution, I focus on the challenges I experienced in tracing and visualizing health data journeys in my empirical exploration of journeys of health data produced in the UK healthcare sector and reused for research purposes conducted between 2018 and 2020.

For the exploration of UK health data journeys, my initial plan was to follow the journeys of patient data flowing to two types of sites: universities and pharmaceutical companies. Therefore, I intended to conduct interviews with key informants at these sites, as well as at data intermediary sites responsible for processing and providing access to patient data for both universities and pharmaceutical companies.

While I successfully recruited key informants at universities, significant challenges arose when attempting to engage with pharmaceutical companies and data intermediary sites. Despite multiple efforts, gaining access to data reuse sites controlled by private-sector actors proved unattainable, mainly due to transparency issues. As a result of these difficulties, the research design was refocused solely on the reuse of patient data for research purposes in universities. This challenge revealed an important feature of the data journeys methodology: its full potential can only be realized when stakeholders involved demonstrate willingness to participate and are transparent. Although access to sites controlled by private-sector actors was not obtained, the process of seeking access provided valuable insights into the characteristics of these sites and showed interesting points of comparison between these sites and sites controlled by different actors (for example, universities).

Furthermore, this access challenge was useful for identifying black boxes of data practices that require the design and application of additional data collection strategies to grey them out or, in other words, to make them less opaque. This highlighted the importance of making adaptations when using this methodology in similar contexts. For instance, where black boxes are identified, it might be necessary to: (1) attend public events to capture the public discourses of people at these sites and key aspects of their culture; (2) invest significant time to try to build contacts with key informants at sites where black boxes are identified, to try to develop an understanding of what is happening at these sites over time; (3) draw on alternative sources to collect data, for example journalistic reports or freedom of information requests.

I also encountered a number of challenges when creating visual representations of data journeys. Through this process, I discovered that visuals are useful to tell the story about the path data follow as they travel between different sites of practice, or in other words, how they move from one point to another. In this sense, the visuals created in the context of this research allowed me to communicate some key insights gained through the exploration of health data journeys regarding how health data are processed and reused within diverse sites of practice.

However, they are less effective in depicting other things that were uncovered through the application of the data journeys methodology, such as how sociocultural values and material factors assemble to shape and justify practices of data reuse; and how in combination these sociocultural values and material factors contribute to the generation of the socio-material conditions that shape data flows. For example, despite multiple attempts, I could not effectively depict how the enthusiasm and excitement of researchers for conducting research with patient data combined with the provision of material resources by funders and other key stakeholders has helped data to flow to the hands of university-based researchers.

Creating effective visuals can be particularly challenging for someone that lacks training as a designer, as was my case. Furthermore, deciding what to include in the visualizations was also a complex task. Visuals can help with explaining tricky concepts and processes that take place within systems (Annan-Callcott, 2021) (for example, how data linkage, aggregation, and de-identification are conducted). However, incorporating representation of these processes into the visualizations may result in overwhelming or convoluted visuals, potentially causing confusion for those engaging with them. Hence, I recommend that researchers who aim to create engaging and effective visualizations of data journeys consider the inclusion of a professional designer in their team. This approach has been successfully adopted by other research teams, as exemplified by the fruitful outcomes of Living with Data (Kennedy et al, 2022).

This contribution focused on the challenges of tracing and visualizing data journeys. These reflections aim to be helpful for others considering the deployment of the data journeys approach in similar contexts.

PART 2: EMPIRICAL CASE STUDIES OF DATA RENDERINGS

Investigating the historical contingencies of a dataset's form – Lindsay Poirier and Quinn White

Datasets structure encoded information by dividing observations into rows, delimiting the variables by which data get described, and grouping like variables through systems of classification. These organizing infrastructures make it possible to sort, aggregate, filter, and plot data. How can data studies scholars make sense of combinatorial social forces through which a dataset's form materializes? How can we make sense of why datasets render certain accounts more visible than others?

As a case study, consider a United States public interest dataset known as Open Payments. In 2013 the US Senate passed the Physician Payments Sunshine Act – legislation that required pharmaceutical and medical device companies to report payments to physicians in this public–facing dataset, which is now published annually by the US Center for Medicare and Medicaid Services (CMS). The legislation passed at a time when 'transparency' was gaining traction as a corporate social responsibility framework and marked a widely held conviction that rendering financial relationships visible to the public could highlight medical conflicts of interests, allow patients to make more informed healthcare decisions, and 'disinfect' biases in prescribing. Today anyone can navigate to a CMS website and look-up how much money their US physicians have accepted from medical manufacturers, along with the nature of the payments. Tabular datasets can also be downloaded from CMS's website for analysis and visualization.

While Open Payments shines a light on financial transactions that have historically taken place behind the scenes, the history of political debates that have shaped its material configurations are less easily discernible. Culturally, datasets are often treated as 'given' – as instrumental tools for analysis, rather than as iterating cultural artefacts. As it turns out, deliberative social action and advocacy can be rendered visible in the dataset, but in order to perceive them an analyst must fix a hermeneutic lens on the dataset's structure and values. This involves curating and interpreting an archive of materials that document the sociocultural provenance of the definitions and standards guiding data reporting.

For example, check out the plot shown in Figure 3.1. It documents how many payments were made to each type of provider in the dataset across two

Figure 3.1: Barplot showing the number of payments to each type of covered recipient in Open Payments across two years

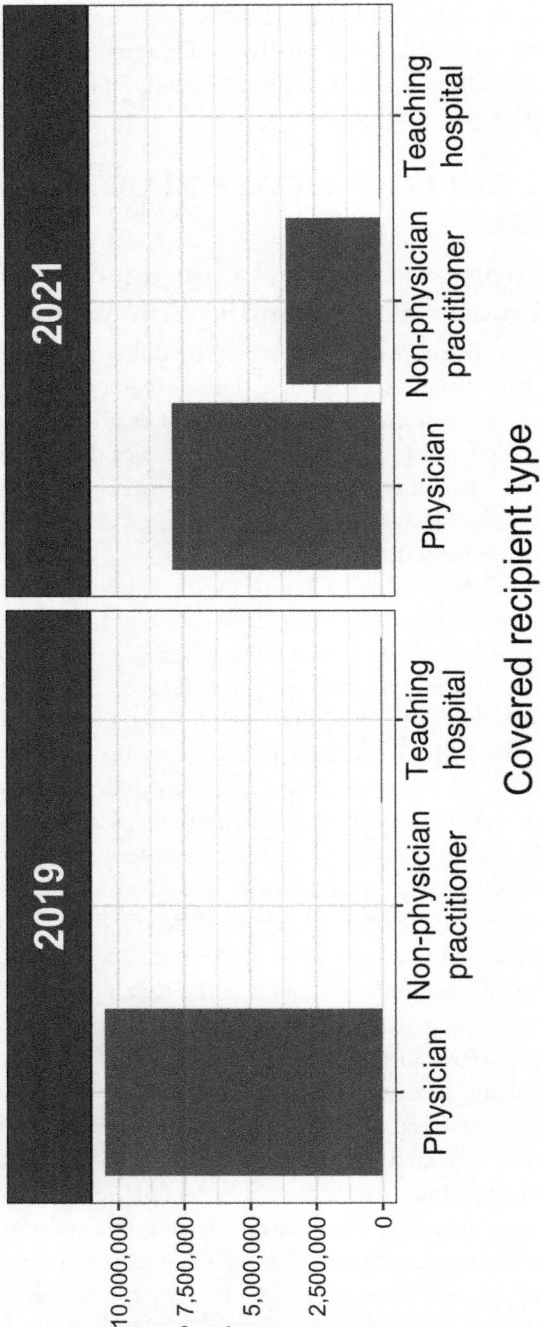

years. We see considerable year-to-year changes in the number of payments made to non-physician prescribers. Up until 2020, manufacturers were only required to report payments made to physicians and teaching hospitals, which notably excluded nurse practitioners, physician assistants, and other non-physician prescribers. We studied the provenance of this decision across several sources – interviewing political and agency staff involved in legislating and standardizing data collection, watching archived videos of senate round tables where the scope of reporting was debated, and reading through hundreds of public comments submitted prior to the formulating of the final standards. We learned that the exclusion of these recipients was contentious. Similar state laws required reporting for these categories of prescribers, and many transparency advocates publicly argued that, in excluding these categories, manufacturers might shift money to non-physician prescribers to evade reporting. In interviews, legislative staff noted that federally legislating reporting for physicians was just a first step, and at the time there were not as many concerns about conflicts of interest when it came to non-physician prescribers. Yet, in the mid-2010s, as the opioid crisis swept the US, there were heightened concerns regarding the role of non-physicians in prescribing opioids. Broad legislation aiming to tackle this epidemic included a provision to expand reporting to these categories of providers.

Political debates can also be seen in the individual values reported in the data. One of the most common lamentations we heard in interviews was how much time was spent while preparing the data standards in 'arguing over bagels'. Imagine this scenario. A pharmaceutical representative shows up to a medical practice and provides a spread of bagels. Throughout the day, physicians pop in and out, grabbing a bagel or two from the spread. Should the pharmaceutical company be required to track the names and consumption of every physician that participated? What if the spread was dropped off at an all-day conference where it may be difficult to oversee who ate what? And does a bagel spread even indicate a potential conflict of interest? In public commentary, industry representatives suggested that having to track this information would place undue regulatory burden on companies. While many activists agreed that the purpose of Open Payments was not to track something as mundane as bagel consumption, they also acknowledged that providing meals was a common way industries attempted to curry favour with physicians and thus needed to be reported accurately. When devising the regulations for the dataset, CMS attempted to find middle ground. When determining the value of a meal to assign to a physician, pharmaceutical and medical device companies would be required to report the 'per person' cost of the meal based on the number of individuals that actually partook, regardless of what they ate or drank. We can see the traces of these arguments and advocacy in visualizations of a company's food and beverage payments to providers on a single date (Figure 3.2).

Figure 3.2: Barplot showing examples where companies reporting to Open Payments paid an identical amount to many providers on a single date

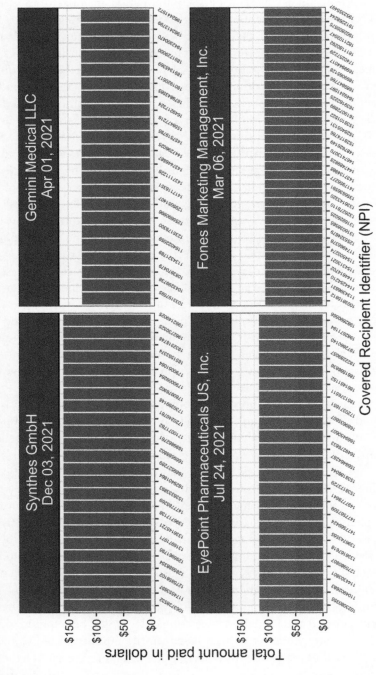

While datasets can bring certain social issues into visibility, their austerity can veil the historical contingencies of the information encoded within them. Fixing a hermeneutic lens on a dataset can draw different data narratives into visibility – narratives that highlight the shifting discursive, regulatory, and sociocultural landscapes from which its form materializes.

Deobfuscating state surveillance in Canada – Evan Light and Ellouise McGeachey

In recent years, there have been calls to regulate state surveillance via strategies such as social movement building and moratoriums on certain technologies to buy policy makers time to make policy. We agree that the situation is urgent and demands an urgent response, yet to plough forward without precision would be a fool's errand. We have observed that it is general practice in Canada for state security agencies to use emerging technologies in their work before relevant legal frameworks have been developed. Today, Canadian state security agencies use a wide range of surveillance technologies with little or no oversight, legal guidance, and accountability. Through our research project, *Deobfuscating State Surveillance Capabilities in Canada*, we seek to document the use of surveillance technologies by state security agencies at every level of government. We do so with a combination of tools: access to information/freedom of information requests, proactive disclosure research, and legal research. Additionally, we maintain all of our research results in a public online archive. This section will map out these tools and how we put them to use.

Government procurement research

There are two general approaches to beginning research on government procurement, both of which we utilize. Each takes a lengthy amount of time, attention to detail, and the ability to manage large volumes of documents.

In the first, we are essentially *lurking* in public-sector procurement venues, documenting procurement in real time by creating accounts on the third-party websites used for government procurement. Sites we find popular with Canadian governments are SAP Ariba (2023),[5] MERX (2023),[6] and Biddingo (2023)[7] and this will vary greatly depending on your jurisdiction. In these spaces, government actors or agencies will issue public requests for goods and services, many of which include finely detailed information. For example, one call we encountered sought bidders to retrofit a Canadian federal prison with a new video surveillance system. The accompanying document included blueprints of the prison, precise notation of the areas to be surveilled and precise descriptions of the types of surveillance to be conducted. Ultimately, it is possible to follow this process into a second wherein correspondence is exchanged between a company and a government actor, a contract is awarded, and work is conducted. Through this process as well as other research, we have developed a living 'Companies of Interest' list, entities whose contracts with governments we believe will help fulfil our goal.

The second method for gathering data is a combination of two processes and is facilitated by open government resources and access-to-information laws. The first (Government of Canada, 2023a) requires the federal government to proactively disclose all contracts of more than $10,000, while the second permits anybody to pay a small sum of money to ask for and receive copies of these contracts, as well as any other government information. For instance, we have uncovered documents showing the Royal Canadian Mounted Police and other federal agencies purchasing forensic management software and cellphone hacking equipment from multiple vendors. In this case, contracts show where equipment has been delivered and in the future we will file further requests to learn precisely how it is used and how its use is governed.

With paltry oversight of policing bodies in Canada, this information will provide us with the information necessary for regulatory and oversight proposals.

Access to information/freedom of information

In Canada, the Access to Information Act[8] facilitates public access to federal government information. One pays $5 to request any number of documents that would be otherwise unpublished. In addition, a live database of previous requests is available and these can be requested for free (Government of Canada, 2023b). Each province and territory in Canada has a Freedom of Information Act which governs how the public can access government information at both provincial and municipal levels. Provinces, territories, and municipalities charge substantial fees both to make requests and fulfil them. When possible, we make requests at the federal level that may provide us with information concerning other levels of government – for instance, email chains, meeting minutes, presentation decks, and briefing notes. Historically a journalistic research technique, practitioners have built their own personal systems of organizing information and their own collections of this information. In an effort to share information and to not duplicate labour, we have created an online archive of our work and will be inviting others to contribute their resources.[9]

Finally, we conduct regular scans of cases within all levels of the Canadian judicial system, seeking cases where surveillance technology has been used in policing. This research enables us to identify, for instance, whether technologies are being used lawfully and if laws and oversight mechanisms exist to control their use. The use of most surveillance technologies in Canada is lightly regulated or entirely unregulated and it is our hope that our research can help lead to a future where policing at all levels is held to a higher level of respect for one's personal privacy.

PART 3: TOOL-BASED INTERVENTIONS

Data Ethics Decision Aid (DEDA) – Petter Falk and Theo Röhle

Data Ethics Decision Aid (DEDA) is an impact assessment workshop developed by the Utrecht Data School (Franzke et al, 2021). The tool functions as a form of deliberative framework specifically designed for use within public administration data projects and is composed of a worksheet with roughly 40 questions on data ethics, project management and organizational responsibilities. These questions are addressed in project teams and working groups in charge of implementing or facilitating data-centred projects. The process provides a critical perspective that outlines (1) the goals of the organization for data–driven tools, (2) the actual results they produce, and (3) the ethical and political aspects that exist between the two.

As a research method, DEDA has been presented as an applied approach for data studies, drawing on a tradition of action research (Schäfer et al, 2022). Such a perspective can contribute with novel or challenging approaches to empirical investigations (Breit et al, 2019; Loukissas, 2019) as well as potentially heading calls for data studies to move towards a more generative critique (Zakharova, 2021). In seeking to address the ethical dimensions of systems at the design stage, rather than at the implementation stage, it shares an agenda with approaches such as 'reflective design' (Sengers et al, 2005) and 'values in design' (Knobel and Bowker 2011). It also ties in with ongoing discussions in computer science that seek to address algorithmic fairness, accountability, and transparency in systems design (Laufer et al, 2022).

Public administration is an area where the relationship between automated and manual decision-making has special urgency. As Gansky and McDonald (2022: 1989) point out: '"Street-level bureaucrats" are often the primary institutionally–provided mechanism for ensuring the realization of normative goals, like fairness, accountability, and transparency – and their discretion can be critical for handling errors, edge cases, and contextual considerations for rule-based systems.' As a research method specifically geared towards this question, DEDA has the potential to highlight very specific real-life settings where the discretionary power of public officers is about to be replaced. By involving both technical and domain expertise, DEDA workshops open up a space for discussing the consequences of such replacements with regard to the values of the organization and the participants. The workshops also create a temporary halt in the developmental pace of the project – a kind of artificially induced 'infrastructural inversion' (Bowker and Star, 2000: 34) – that renders aspects visible that otherwise tend to fade into the background of taken-for-granted 'enabling environments' (Peters, 2015: 3) of data infrastructures.

As such, the methodology has the potential to practically challenge notions of data objectivity and address power imbalances inherent in data practices and algorithmic systems, while the functionality of the system is not yet established. Furthermore, there is an opportunity to build on the growing body of empirical material exploring DEDA as an applied critical research methodology as the basis for a larger comparative outlook. Originating in the Netherlands, the framework has been used in over 100 workshop settings in Dutch municipalities and government agencies, starting in 2016. As of 2023, the DEDA method has been applied in a handful of research and development settings across Europe. Researchers at Karlstad University and Gothenburg University have adapted the method for the Swedish context, and research collaborations between the Netherlands, Germany, Sweden, Finland, and Estonia are under way. There are a number of challenges for a comparative project of this scale, but mapping similarities and differences between various local data settings holds a significant potential to gauge the importance of ethical questions in an inductive manner, rather than simply applying a given ethical framework.

Sphere Transgression Watch: reflections on translating a philosophical theory to a digital tool – Marthe Stevens, Tamar Sharon, and Bernard van Gastel

In the past decade, the large tech corporations which we know so well from the world of computational software, hardware, and the Internet, have expanded well beyond their original sphere of activity into new societal domains, or 'spheres'. For example, major tech companies have been developing software and wearables for remote clinical studies (Apple, 2015), making their mark with e-learning platforms which seek to personalize learning and facilitate remote teaching (Google, 2014; Apple, 2018), and assisting farmers with sustainable and regenerative agriculture practices (X.company, 2020).

In order to study this Big Tech expansionism, we developed the Sphere Transgressions theoretical lens, drawing on Michael Walzer's theory of justice (1983). In short, Walzer argues that social life is made up of different spheres that are organized around different conceptions of justice. In a just society, advantages in one sphere – such as wealth or political power – should not translate into advantages in another. Such translations consist in 'sphere transgressions' and ultimately unjust and tyrannical societies. We argue that we are currently witnessing a series of sphere transgressions by tech companies, whereby the (legitimate) advantages they have accrued in the sphere of digital goods, namely digital expertise and know-how, are translated into (illegitimate) advantages in other societal spheres (Sharon, 2021a, 2021b; Stevens et al, 2024).

While studying this growing phenomenon, we learned that Big Tech corporations were expanding quickly and in societal spheres well beyond our main area of expertise (health and medicine). Also, we noticed that there was little awareness about this phenomenon in non-academic settings. Therefore, we decided to translate the Sphere Transgressions theoretical lens into visual form. We gathered a team of philosophers, social scientists, computer scientists, and interaction designers all working at Radboud University's interdisciplinary Hub for Digitalization and Society (iHub, the Netherlands). Collectively, we were able to launch the Sphere Transgression Watch (STW), a digital tool that tracks the growing involvement of eight large tech corporations in various societal sectors, in April 2022 (Stevens et al, 2022). It is a public-facing, open data tool, which seeks to both visualize and render more tangible the threat of Big Tech expansionism through society for a broad audience. Users of the tool can scroll through the timeline and see the influence of tech corporations in particular societal spheres growing. They can also click on a particular sphere and gain access to a rich archive of newspaper articles, blogs, and information from company websites that can be used for research purposes.

Translating a theoretical lens into visual form was not straightforward and we had to navigate many questions related to in/visibility and power. One of the first questions that we had to answer was how to chart the contours of something that is 'invisible'. We brainstormed about ways to collect relevant data for the tool as there was no public archive where all Big Tech's initiatives are registered. We realized that we should collect the data ourselves and that the best sources were newspaper articles, blogs, and information from company websites that were used to present, promote, or reflect on Big Tech's innovations. But simply listing these – often very positive – messages could easily be seen as a celebration of Big Tech's innovative capabilities, which was certainly not our intention.

In order to render more tangible the threat of Big Tech expansionism through society, we had to ask ourselves how we should visualize sphere transgressions and what aspects to foreground in the visualization. Figures 3.3 and 3.4 show drawings made during our brainstorming sessions. Figure 3.3 focuses on companies and how they expand to various societal domains. In contrast, Figure 3.4 foregrounds societal spheres and how they are being infiltrated by large tech corporations. In the end, we decided to focus on societal spheres and how the influence in the spheres grows over time (Figure 3.5) as we wanted to highlight the extent of this phenomenon in various societal spheres at the same time. These decisions made us realize how seemingly small design choices had an important impact on the visualization and how it would be perceived by others. For example, in selecting the colours for the website, we opted for

Figure 3.3: Drawing that shows how Big Tech expands to various societal spheres

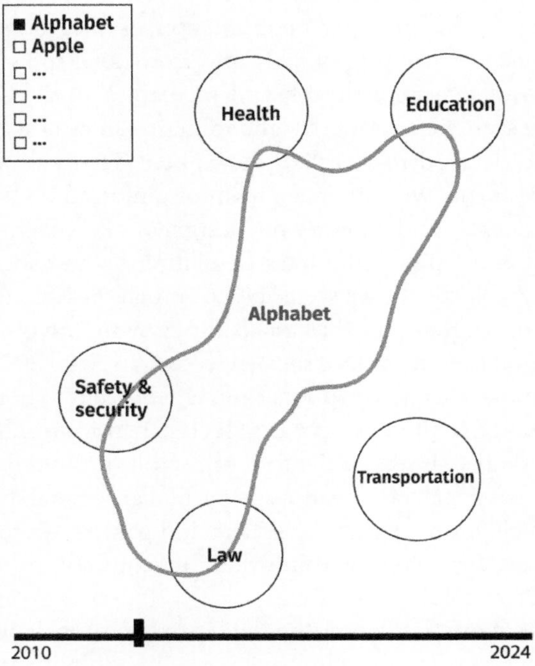

a colour scheme which suggests risks and harms: red, black, and white, rather than more reassuring colours such as green or blue.

There is value in translating a philosophical theory to a digital tool. We noticed that the development of visualizations can be an important way to refine a theoretical lens. Theoretical concepts needed to be made concrete and so needed to be defined early on. For example, the technical experts needed to know how many spheres should be included on the website. Or, could an initiative fall into multiple spheres at the same time? Are certain initiatives more significant than others? Such questions were difficult for the philosophers, who preferred not to see spheres as well-defined entities, but as a heuristic that can be defined on various levels depending on their use. Eventually the philosophers were able to pin-down some concepts for the website, but also argued for a highly flexible tool to which they could make alterations later on. For example, they wanted the possibility to change the names of the spheres, if one turned out to be problematic later on. Looking back, there was value in defining such concepts early on, as this forced the team to think through the consequences of theoretical choices earlier.

In addition, we noticed that the visualizations allowed our theoretical ideas to travel further and reach new audiences. For example, the website is

Figure 3.4: Drawing that shows how societal spheres are infiltrated by Big Tech initiatives

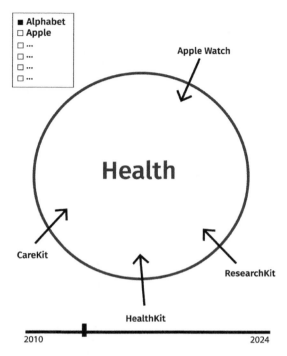

currently being used for research purposes and to bring researchers together for a special issue (Stevens et al, 2024). It is also used in presentations that members of the research group give in non-academic settings, such as policy circles (Dutch Ministry of the Interior) and professional conferences (for example, on AI in health for medical professionals). The tool has been highlighted in Dutch and international media (for example, BNR Radio, De Volkskrant, and the Swiss newspaper *Le Temps*) thereby stimulating timely discussions about Big Tech expansionism.

Conclusion

The contributions of this chapter highlight how the themes of visibility and invisibility offer a central moment of reflection on how we can and ought to study and understand digital data in our research. They highlight the challenges and issues, but also the opportunities that come with questions of positionality, access, transparency, as well as questions of infrastructure and bureaucracy. This enables us to see – and as such shows us – that the kaleidoscope does more than provide different perspectives. While each

Figure 3.5: The final design of the digital tool

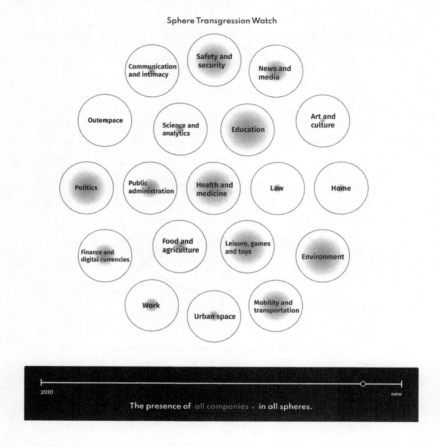

of the contributions highlights different aspects of power and different subjects of study in data studies, the contributions are bound together by the kaleidoscope itself.

The kaleidoscope enables multiplicity in a way that enables us to relate these perspectives *to each other*. Similarly, the contributions of this chapter show different perspectives on (methodological) challenges in data studies but are bound together in a common understanding of data studies that does not only accept multiplicity, but ties it together in a shared world that is related and multiple at the same time – or as John Law (2002) argues: fractal. The question then is not only what is being made visible in the different perspectives, but how can these be brought together without denying that they are fractal enactments? In this sense, our choice of the kaleidoscope as metaphor addresses the dilemma that Vogl (2007: 22) ascribes to the historical development of the telescope from instrument to medium: 'every visibility is surrounded by an ocean of invisibility. ... With every deepening of clarity comes a new depth of the unclarifiable.'

Earlier in the chapter, Zakharova argues that what we know about processes of datafication of society relies to a large degree on the methods and approaches that we are using. Thus, what is being made visible (in an always already structured world) relies in a first moment on the modes of engagement that we choose. In an interdisciplinary field like data studies, there is the necessity to become open and transparent to others but also to oneself about what part of the kaleidoscope we are using to look at the complex and multiple phenomenon of datafication.

And this points towards the need to also make ourselves visible in our encounters and our research. This choice in itself is already political, as it results from a theoretical contingency that needs to be translated into a practical singular perspective. What do we observe? What do we measure? And what is being observed and made in/visible in the datasets that we work with? Poirier and White show how a political act for more transparency in Open Payment initiatives is not only qua existence political, but also in the nitty-gritty details of its design. What about the bagels? Should we track how the industry provides food for US physicians? This example demonstrates that activists themselves are not always clear how to respond to that question. The contribution of the kaleidoscope of data studies here is to show that data as such are not sufficient to address the political dimension of datafication, but that we also need to be critical about the nuanced design questions of digital infrastructures that construct, transport, and translate data. This does not only include questions of databases, servers, or network switches, but also refers to the fact that these infrastructures are embedded and part of an (always already existing) organizational structure that runs them, works with them, and makes sense of them.

As Light and McGeachey show, organizational (presentations, meeting notes) and legally mandated practices (public tenders) create traces that can enrich our perspective on the datafication of society. On the same note, these infrastructures and the data practices they are co-shaping can become visible upon breakdown. As Fahimi et al argue, infrastructural breakdowns are important moments of revealing how these organizations work and how data are a part of their operational logic. Moments of breakdown also offer moments of reflection and opportunities for intervention. Falk and Röhle introduce with DEDA a tool-based intervention that provokes moments of infrastructural inversion (Bowker and Star, 2000) – and thus create calculated moments of possible change – and with that adding a facet to the kaleidoscope where not only the perspectives become multiple, but also the very phenomenon that we study. This, of course, also means that we as scholars and activists have even more responsibility: involvement in translations is also political engagement. Politics here means that it could always be otherwise[10] (Mol, 2002; Latour, 2004). This is reflected in the question that Stevens et al were facing: how do we visualize sphere transgressions to make the threat more tangible and how do we avoid being a proponent of Big Tech? Or in other words: how do we translate our inquiry in a way that makes aspects visible that are abstract and based in a theoretical potentiality? It is also reflected in the challenges experienced by Medina Perea, for example, how to trace and visualize the movement of data when gaining access to data reuse sites controlled by private-sector actors is unattainable, mainly due to transparency issues.

Coming back to the starting metaphor of the kaleidoscope. In our opinion, the field of data studies is the kaleidoscope that enables multiple perspectives, addresses different forms of in/visibilities and thematizes the political dimension of these methodological inquiries. Considering distinct perspectives, as highlighted in this chapter, allows us to gain deeper insights into the social consequences of data. It is thereby clear that the field, as interdisciplinary as it is, by itself constitutes a multiplicity, and that the multiple engagements, translations, and interventions multiply this even more. In these engagements, we add new perspectives, lose others, and thus change the kaleidoscope as such, allowing at any given moment for contingency and fluidity. However, what holds the kaleidoscope together is the insight that our methodological approaches and the in/visibilities they create and tackle at the same time are born out of an understanding that goes beyond a modernist narrative of data as description of a detached reality (see also Bowker, 2006; Gitelman, 2013). As such, the metaphor of the kaleidoscope is the always present reminder that our methodological engagements are situated, reflexive, and political.

Discussant – Jonathan W. Y. Gray

How to study what data makes (in)visible and with what consequences? The sections in this chapter explore methods, approaches, tools, and interventions for accounting for the social lives of data in society – from tracing frictions and fixes, to following journeys and histories, to making freedom of information requests, maps, and visualizations. In being invited to serve as a discussant for this chapter and to surface connections with my own work, I ruminate on resonances with shared research practices for studying data – the details of what is done, what is asked, what is noticed – as well as on how different ways of accounting for data can make a difference.

As alluded to throughout the chapter, feminist science and technology studies scholars emphasize situated and relational approaches – with particular attentiveness to who and what is missing, who and what might be affected by but absent from the shaping of technoscience. Susan Leigh Star wrote memorably of her dedication to studying relations between 'lived experiences', 'technologies', and 'silences' (Star, 2007). Donna Haraway encourages 'staying with the trouble', 'nurturing capacities to respond', and 'cultivating ways to render each other capable' (Haraway, 2016). Feminist data scholars also emphasize ethics of care (Taylor, 2020; Fotopoulou, 2021), the possibilities and politics of refusing data (Barabas, 2022; Garcia et al, 2022), and affirming epistemic plurality, counter-narratives, and marginalized perspectives (D'Ignazio and Klein, 2020).

These approaches may provide methodological inspiration and orientation in studying data. In a recent project studying open data portals, which may be considered devices for making data public, data request mechanisms provide an indication of datasets which were sought but not found (Gray, 2023).[11] Just as Mimi Ọnụọha's Library of Missing Datasets provides a 'repository of those things that have been excluded in a society where so much is collected',[12] so gathering unsuccessful requests from portals around the world may provide an indication of data considered missing, users left disappointed, questions unanswered, and how data practices, policies, and portals have failed to live up to aspirations and expectations. As well as disclosing details about data infrastructures, data portal interfaces may disclose the failure of arrangements meant to inform. Attending to and articulating trouble has become part of digital culture and society (Meunier et al, 2021), as well as a way to situate and reflect on taken-for-granted digital practices (van Geenen et al, 2023).

How might data studies modify data practices? How can methodological enactments for understanding and situating data (whether through tracing data journeys and frictions, maps, or information requests) make a difference to its role in culture, society? As part of a collaboration between those who do and study data journalism, we identified themes and challenges for 'critical data practices' (Bounegru and Gray, 2021). For example, drawing on a chapter by Helen Verran (2021) on narrating numbers and staying with the trouble of value, we considered how data journalists might tell stories both with and about data – that is, both using data as a medium for exploring and narrating issues, as well as reporting on the making of data.

Reporting on data (in)visibilities may be relevant for both those using and those studying data in society – as illustrated by a 2021 *Washington Post* piece exploring what carbon emissions data suggests about both the emission of carbon as well as the making of data.[13] Such approaches to following data in society may suggest a shift from looking at datasets as resources, to exploring how data infrastructures embody relations between actors, organizations, processes, and methods – and how other kinds of datasets may be made and other arrangements may be composed (Gray et al, 2018).

The more-than-academic relevance of investigating data infrastructures and data (in)visibilities may also surface through collaborations which could be characterized as inventive or experimental (Lury and Wakeford, 2012; Lezaun et al, 2016). As well as critically reconsidering conventional infrastructures and outputs of scholarly communication – from societies to monographs to pay-walled peer reviewed journal articles (Eve and Gray, 2020) – the modification and hybridization of formats may provide generative entrypoints for exploring the broader salience of data studies (de Mourat et al, 2020). Careful attention to how questions are formulated, problems are articulated, methods are enacted, and communities are assembled may enrich collaborative investigations with and about data (Gray et al, 2022). Thinking along with feminist science and technology studies, the kinds of methods, tools, and approaches outlined in this chapter may contribute not only to the study of data in society (including scrutinizing its role in the social production of invisibilities, absence, and marginalization) but also to its unscripting, taking-apart, respecification, repurposing, and recomposition.

Notes

1. Alphabetical author order
2. Discussant
3. Facilitator
4. Research presented in this section draws on Irina's doctoral thesis which she received from the University of Bremen.
5. Available at: https://service.ariba.com
6. Available at: www.merx.com/
7. Available at: www.biddingo.com
8. Available at: https://laws-lois.justice.gc.ca/eng/acts/A-1/
9. Available at: https://surveillance.glendon.yorku.ca/
10. A great and very fitting variation of this theme – namely: 'IT could be otherwise' – has been developed by the Technologies in Practice Group (TiP) at the IT University Copenhagen.
11. https://datanotfound.jwyg.org/
12. https://mimionuoha.com/the-library-of-missing-datasets
13. https://www.washingtonpost.com/climate-environment/interactive/2021/greenhouse-gas-emissions-pledges-data/

References

Annan-Callcott, G. (2021, 2 September) *What we know about communicating how health data is used | Understanding patient data*. Understanding Patient Data. Available at: http://understandingpatientdata.org.uk/

Apple (2015, 9 March) 'Apple introduces ResearchKit, giving medical researchers the tools to revolutionize medical studies'. *Apple Newsroom*. Available at: www.apple.com/newsroom/2015/03/09Apple-Introduces-ResearchKit-Giving-Medical-Researchers-the-Tools-to-Revolutionize-Medical-Studies/

Apple (2018, 26 June) 'Apple's free Schoolwork app now available for teachers'. *Apple Newsroom*. Available at: www.apple.com/newsroom/2018/06/apples-free-schoolwork-app-now-available-for-teachers/

Barabas, C. (2022) Refusal in data ethics: Re-imagining the code beneath the code of computation in the carceral state. *Engaging Science, Technology, and Society*, 8: 35–57.

Barad, K. (2007) *Meeting the Universe Halfway: Quantum physics and the entanglement of matter and meaning*. Duke University Press.

Bates, J., Kennedy, H., Medina Perea, I., Oman, S., and Pinney, L. (2023) Socially meaningful transparency in data-based systems: Reflections and proposals from practice. *Journal of Documentation* (ahead of print). Available at: https://doi.org/10.1108/JD-01-2023-0006

Bates, J., Lin, Y.-W., and Goodale, P. (2016) Data journeys: Capturing the socio-material constitution of data objects and flows. *Big Data & Society*, 3(2). Available at: https://doi.org/10.1177/2053951716654502

Bhaskar, R. (2008) *A Realist Theory of Science*. Routledge.

Bounegru, L. and Gray, J. (eds). (2021) *The Data Journalism Handbook: Towards a critical data practice*. Amsterdam University Press.

Bowker, G. C. (2006) *Memory Practices in the Sciences*. MIT Press.

Bowker, G. C. and Star, S. L. (2000) *Sorting Things Out: Classification and its consequences*. MIT Press.

Breit, E., Egeland, C., and Løberg, I. B. (2019) 'Cyborg Bureaucracy: Frontline Work in Digitalized Labor and Welfare Services'. In: J. S. Pedersen and A. Wilkinson (eds) *Big Data*. Edward Elgar Publishing, pp 149–169.

Burrell, J. (2016) How the machine 'thinks': Understanding opacity in machine learning algorithms. *Big Data & Society*, 3(1). Available at: https://doi.org/10.1177/2053951715622512

Christin, A. (2020) The ethnographer and the algorithm: Beyond the black box. *Theory and Society*, 49(5): 897–918. Available at: https://doi.org/10.1007/s11186-020-09411-3

D'Ignazio, C. and Klein, L. (2020) *Data Feminism*. MIT Press.

de Mourat, R., Ricci, D., and Latour, B. (2020) 'How Does a Format Make a Public?' In M. P. Eve and J. Gray (eds) *Reassembling Scholarly Communications: Histories, infrastructures, and global politics of open access*. MIT Press.

Dencik, L., Hintz, A., Redden, J., and Treré, E. (2022) *Data Justice*. Sage.

Eve, M. P. and Gray, J. (eds) (2020) *Reassembling Scholarly Communications: Histories, infrastructures, and global politics of open access*. MIT Press.

Fotopoulou, A. (2021) Conceptualising critical data literacies for civil society organisations: Agency, care, and social responsibility. *Information, Communication & Society*, 24: 1640–1657.

Franzke, A. S., Muis, I., and Schäfer, M. T. (2021) Data Ethics Decision Aid (DEDA): A dialogical framework for ethical inquiry of AI and data projects in the Netherlands. *Ethics and Information Technology*, 23(3): 551–567.

Gansky, B. and McDonald, S. (2022) 'CounterFAccTual: How FAccT Undermines Its Organizing Principles', *2022 ACM Conference on Fairness, Accountability, and Transparency*, pp 1982–1992.

Garcia, P., Sutherland, T., Salehi, N., Cifor, M., and Singh, A. (2022) No! Re-imagining data practices through the lens of critical refusal. *Proceedings of the ACM on Human–Computer Interaction*, 6: 315:1–315:20.

Gitelman, L. (2013) *"Raw Data" Is an Oxymoron*. MIT Press.

Google (2014, 6 May) 'Previewing a new Classroom'. *Google blog*. Available at: https://blog.google/outreach-initiatives/education/previewing-new-classroom/

Government of Canada (2023a) *Proactive Disclosure*. Available at: https://open.canada.ca/en/proactive-disclosure

Government of Canada (2023b) *Completed Access to Information Requests*. Available at: https://open.canada.ca/en/search/ati

Graham, S. and Thrift, N. (2007) Out of order: Understanding repair and maintenance. *Theory, Culture & Society*, 24(3): 1–25. Available at: https://doi.org/10.1177/0263276407075954

Gray, J. W. Y. (2023) What do data portals do? Tracing the politics of online devices for making data public. *Data & Policy*, 5, e10.

Gray, J., Gerlitz, C., and Bounegru, L. (2018) Data infrastructure literacy. *Big Data & Society*, 5: 1–13.

Gray, J., Bounegru, L., Rogers, R., Venturini, T., Ricci, D., Meunier, A., et al (2022) Engaged research-led teaching: Composing collective inquiry with digital methods and data. *Digital Culture & Education*, 14: 3.

Haraway, D. J. (2016) *Staying with the Trouble: Making kin in the Chthulucene*. Duke University Press Books.

Katzenbach, C. (2021) 'AI will fix this': The technical, discursive, and political turn to AI in governing communication. *Big Data & Society*, 8(2). Available at: https://doi.org/10.1177/20539517211046182

Kennedy, H., Oman, S., Ditchfield, H., Taylor, M., Bates, J., Medina Perea, I., et al (2022) *Data Matters Are Human Matters: Final Living With Data report on public perceptions of public sector data uses*. The University of Sheffield. Available at: https://livingwithdata.org/project/wp-content/uploads/2022/10/LivingWithData-end-of-project-report-24Oct2022.pdf

Kitchin, R. (2017) Thinking critically about and researching algorithms. *Information, Communication & Society*, 20(1): 14–29. Available at: https://doi.org/10.1080/1369118X.2016.1154087

Knobel, C. and Bowker, G. C. (2011) Values in design. *Communications of the ACM*, 54(7): 26–28. Available at: https://doi.org/10.1145/1965724.1965735

Latour, B. (2004) Whose cosmos, which cosmopolitics? Comments on the peace terms of Ulrich Beck. *Common Knowledge*, 10(3): 450–462.

Laufer, B., Jain, S., Cooper, A. F., Kleinberg, J., and Heidari, H. (2022) 'Four Years of FAccT: A Reflexive, Mixed-Methods Analysis of Research Contributions, Shortcomings, and Future Prospects', *2022 ACM Conference on Fairness, Accountability, and Transparency*, pp 401–426.

Law, J. (2002) *Aircraft Stories: Decentering the object in technoscience* (1st edn). Duke University Press.

Law, J. (2004) *After Method: Mess in social science research.* Routledge.

Law, J., Ruppert, E., and Savage, M. (2011) *The double social life of methods.* Centre for Research on Socio-Cultural Change (CRESC), Working Paper No 95.

Lezaun, J., Marres, N., and Tironi, M. (2016) 'Experiments in Participation'. In U. Felt, R. Fouche, C. A. Miller, and L. Smitt-Doer (eds) *Handbook of Science and Technology Studies* (4th edn). MIT Press, pp 195–222.

Loukissas, Y. A. (2019) *All Data Are Local: Thinking critically in a data-driven society.* MIT Press.

Lury, C. and Wakeford, N. (eds). (2012) *Inventive Methods: The happening of the social.* Routledge.

Meunier, A., Gray, J., and Ricci, D. (2021) 'A New AI Lexicon: Algorithm Trouble: Troublesome Encounters with Algorithms that Go Beyond Computational Processes'. In *A New AI Lexicon*. AI Now Institute.

Mol, A. (2002) *The Body Multiple: Ontology in medical practice.* Duke University Press.

Peters, J. D. (2015) *The Marvelous Clouds: Toward a philosophy of elemental media.* University of Chicago Press.

Powell, A. B., Ustek-Spilda, F., Lehuedé, S., and Shklovski, I. (2022) Addressing ethical gaps in 'Technology for Good': Foregrounding care and capabilities. *Big Data & Society*, 9(2): 20539517221113776. Available at: https://doi.org/10.1177/20539517221113774

Ruppert, E., Law, J., and Savage, M. (2013) Reassembling social science methods: The challenge of digital devices. *Theory, Culture & Society*, 30(4): 22–46. Available at: https://doi.org/10.1177/0263276413484941

Schäfer, M., Falk, P., Muis, I., and Röhle, T. (2022) 'Analysing Data Power through Participatory Observation: A Call to Action', Data Power 2022 Conference.

Seaver, N. (2017) Algorithms as culture: Some tactics for the ethnography of algorithmic systems. *Big Data & Society*, 4(2): 2053951717738104. Available at: https://doi.org/10.1177/2053951717738104

Sengers, P., Boehner, K., David, S., and Kaye, J. (2005) 'Reflective Design', *Proceedings of the 4th Decennial Conference on Critical Computing: Between Sense and Sensibility*, pp 49–58.

Sharon, T. (2021a) Blind-sided by privacy? Digital contact tracing, the Apple/Google API and big tech's newfound role as global health policy makers. *Ethics and Information Technology*, 23: 45–57. Available at: https://doi.org/10.1007/s10676-020-09547-x.

Sharon, T. (2021b) From hostile worlds to multiple spheres: Towards a normative pragmatics of justice for the Googlization of health. *Medicine, Health Care & Philosophy*, 24: 315–327. Available at: https://doi.org/10.1007/s11019-021-10006-7.

Star, S. L. (1999) The ethnography of infrastructure. *American Behavioral Scientist*, 43(3): 377–391. Available at: https://doi.org/10.1177/00027649921955326

Star, S. L. (2007) 'Five Questions'. In J. K. B. Olsen and E. Selinger (eds) *Philosophy of Technology*. Automatic Press, pp 227–233.

Star, S. L. and Ruhleder, K. (1996) Steps toward an ecology of infrastructure: Design and access for large information spaces. *Information Systems Research*, 7(1): 111–134. Available at: https://doi.org/10.1287/isre.7.1.111

Stevens, M., Kraaijveld, S. R., and Sharon, T. (2024) Sphere transgressions: Reflecting on the risks of big tech expansionism. *Information Communication & Society*, 1–14. Available at: https://doi.org/10.1080/1369118X.2024.2353782

Stevens, M., Sharon, T., van Gastel, B., Hoffman, A. S., Kraaijeveld, S. R., and Siffels, L. (2022) *Sphere Transgression Watch*. Distributed by iHub, Available at: www.sphere-transgression-watch.org

Swist, T., Collin, P., and Third, A. (2019) Children's data journeys beyond the 'supply chain': Co-curating flows and frictions. *Media International Australia*. https://doi.org/10.1177/1329878X19828390

Taylor, L. (2020) The price of certainty: How the politics of pandemic data demand an ethics of care. *Big Data & Society*, 7: 2053951720942539.

van Geenen, D., van Es, K., and Gray, J. W. Y. (2023) Pluralising critical technical practice. *Convergence*, 30(1): 7–28.

Verran, H. (2021) 'Narrating a Number and Staying With the Trouble of Value'. In L. Bounegru and J. Gray (eds) *The Data Journalism Handbook*. Amsterdam University Press, pp 55–64.

Vogl, J. (2007) Becoming-media: Galileo's telescope. *Grey Room*, 29: 14–25.

Walzer, M. (1983) *Spheres of Justice: A defense of pluralism and equality*. Basic Books.

X.company (2020, 12 October) 'Mineral: Bringing the era of computational agriculture to life'. *X.company blog*. Available at: https://blog.x.company/mineral-bringing-the-era-of-computational-agriculture-to-life-427bca6bd56a

Zakharova, I. (2021) Data inquiry: Methodological considerations on datafication in social research. *AoIR Selected Papers of Internet Research.* https:// doi.org/10.5210/spir.v2021i0.12082

Zakharova, I. (2022) *Understanding data studies: A conceptual and methodological inquiry into research on datafication.* PhD thesis, Universität Bremen. Available at: https://doi.org/10.26092/elib/1675

Ziewitz, M. (2016) Governing algorithms: Myth, mess, and methods. *Science, Technology, & Human Values*, 41(1): 3–16. Available at: https://doi.org/ 10.1177/0162243915608948

4

People's Practices in the Face of Data Power

Roderic Crooks,[1] Catherine D'Ignazio,[2] Arne Hintz,
Fieke Jansen, Juliane Jarke,[3] Anne Kaun, Stine Lomborg,
Dan McQuillan,[4] Jonathan A. Obar, Lucy Pei, and Ana Pop Stefanija

Introduction

The shift towards big data-driven decision-making and algorithmic automation across many aspects of everyday life remains a contentious subject of debate and critique. Critical social scientists and media scholars assert that this shift alters the nexus and power relations between state, citizens, and industry (for example, Kennedy and Moss, 2015). The extractive logic central to today's data economy has further centralized power, wealth, and capital in the hands of the few industry leaders (Srnicek, 2017; Cohen, 2019). The desire of states to reap the perceived benefits of data use for optimization, efficiency, and control is increasing the use of (commercial) data systems. Individuals and communities have little control over how their data are collected and have little to no influence on the algorithmically informed decisions that govern their lives. We will refer to this power asymmetry as a difference in data power.

However, this data power does not affect everyone equally, and some people are more resourceful in (temporarily) pushing back against or working around processes of datafication. In that sense, we need to ask how these data and automated decision-making processes are shifting power, how they work, for whom and for whom not. Members of marginalized, racialized, and vulnerable communities experience the brunt of data power. For example, research has shown such communities to be a target of automated decisions within welfare states, such as being more likely to be subjected to algorithmic fraud detection in social services and datafied policing (Eubanks, 2017; Roosen, 2020; Jansen, 2022). Yet people and communities are challenging and negotiating the influence and impact of this hegemonic data power. This chapter will highlight a range of practices developed by people in the face of data power.

The approaches we discuss here range from individual to community-based and collective practices that push back against state power as well as the power exerted by private companies and interests. The progression from the individual towards the collective, from the private towards the public, captures the underlying rationale that individual acts of reclaiming data power are necessary but need to be complemented by collective approaches in order to address the wider implications and transformations of increasingly datafied societies. The practices we outline and envision in this chapter exemplify these different dimensions, although they do not aim to be exhaustive of the power people enact through their data practices. They demonstrate what is possible, capturing a range of possibilities and potential incentives for reimagining, reclaiming, and building better datafied futures.

The practices brought to the fore are meaningful consent, refusal as an act of agency, data literacy and collective agency, community activism, and participatory governance approaches. Challenges associated with realizing meaningful forms of consent are presented as a central set of concerns that call for further privacy research and advocacy to support individuals as they attempt to realize data power. If the embedding of the meaningful consent mechanism through regulatory and technological means is an essential first step for empowering individuals, enabling refusal practices as an act of control and agency is the next. The process of refusal is about more than saying no, it is about the willingness, knowledge, and ability to exercise refusal. As an act of 'speaking back' and shifting power, it enables people to act according to their own will. Yet it requires the willingness and ability to invest time and energy, and therefore, as a practice, faces limitations.

People's practices might provide important indications to the kinds of structures needed to support their communal and collective exercise of power in relation to state and commercial actors. Challenging the use of algorithmic welfare systems and exercising redress in the face of data power cannot be the sole responsibility of the individuals or communities impacted. It requires a mix of skills, knowledge, and voice(s), to mobilize what is referred to as collective agency. One such example is community data activism as a new vector for participatory power. Participatory governance approaches are another emerging and possible practice of opposing state-exerted data power. Through democratic innovations such as citizen assemblies, citizen juries, and others, citizen voices are advanced in decision-making processes. Focusing on giving voice to those impacted the most by these power apparatuses, these methods allow citizens to partake in complex conversation and are seen as positive expressions of agency, although they do not transfer decision-making power to the citizen, yet. While these emerging practices might seem marginal in the face of data power, they can give people and communities knowledge and voice in datafied societies.

Meaningful consent and data power – Jonathan Obar

The data subject oversees their own information protections. This assertion is fundamental to privacy law and regulation based in the Fair Information Practice Principles (Cate, 2006). The European Union's General Data Protection Regulation (GDPR) and Canada's Personal Information Protection and Electronic Documents Act (PIPEDA) are two international examples of this approach. Both emphasize the importance of the data subject in the realization of their own information protections, conveyed via the centrality of consent provisions. These provisions are supposed to afford opportunities for assenting (or dissenting) to the implications of service engagement, aligned with calls for 'democratis(ing) data power' (Kennedy and Moss, 2015: 2). Meaningful consent suggests that individuals not only engage with consent materials before deciding whether to agree or not, but that individuals also understand what they are agreeing to, as well as the implications of agreement (OPC, 2020). Those implications might be that data collected today are integrated into artificial intelligence (AI) development in the future. If consent processes shift data power towards the individual, meaningful dissent expressions might result, with individuals refusing to consent to AI development possibilities. Unfortunately, current methods for delivering meaningful forms of online consent suggest a difficult set of challenges for realizing this type of information protection.

'I agree to the terms and conditions' is said to be 'the biggest lie on the internet' (Lannerö, 2012). Research suggests individuals tend to ignore terms of service and privacy policies when clicking 'agree' during app/ website sign-up (Obar and Oeldorf-Hirsch, 2020, 2022). In two empirical studies, one of undergraduates (n=543), and another of older adults 50+ (n=500), participants were presented with the front page of a fake social network (called NameDrop) and asked to engage with a fictitious sign-up and associated consent process (Obar and Oeldorf-Hirsch, 2020, 2022). In both studies the majority of participants agreed to the fake privacy policy without accessing or reading it via a form of clickwrap. For those that accessed the text of the policies, average and median reading times suggested that many participants were likely scrolling to the bottom of the policy as quickly as possible without reading or understanding service terms. To assess the implications of potential policy-ignoring behaviours, in both studies 'gotcha clauses' were included in the terms of service. In the undergraduate study, 93 per cent agreed to give NameDrop their first-born child for service access. In the older adult study, 83 per cent agreed to give a kidney or other bodily organ (Obar and Oeldorf-Hirsch, 2020, 2022). These findings suggest that consent provisions, as currently presented to data subjects during service sign-up in particular, are not delivering information protections.

There are many reasons data subjects struggle with online consent processes. The length and complexity of service terms are a longstanding concern (McDonald and Cranor, 2008; Reidenberg et al, 2015; Obar, 2022a, 2022b). The problematic user interface designs of digital services can also make it difficult to realize information protections (Acquisti et al, 2017). The literature suggests the deceptive design of an online consent process can distract and even discourage people from engagement and understanding of service terms (Obar and Oeldorf-Hirsch, 2018; Habib et al, 2020). Due to these difficulties, it is no surprise that data subjects may also deal with feelings of resignation and apathy when attempting to realize information protections (Hargittai and Marwick, 2016; Draper and Turow, 2019).

How to achieve a consistent and meaningful consent across myriad consent scenarios online is a question without a clear answer. There are a variety of approaches that are part of the ongoing discourse attempting to support consent processes. This includes efforts to encourage service provider self-regulation in the form of consent processes that are more dynamic, as opposed to obtaining consent during service sign-up or when policies change (OPC, 2021). The 'just-in-time' notice (OPC, 2021) is an example of this type of ongoing consent process, which would alert individuals to new opportunities for considering whether to consent or not, associated with specific online behaviours such as turning on a device camera or posting content online. To encourage better self-regulation by digital service providers, policy makers internationally are imposing monetary penalties where the design of online consent interfaces suggests organizations are maintaining unhelpful and even deceptive practices (Obar, 2023). Strategies for making notice materials more engaging continue to be tested. This includes calls for policies with language and formatting that are easier to understand (OPC, 2021), and supplementary services that distil complex details (TOS;DR, 2023). Nutrition label-type notice materials, along with thoughtful signage examples are also being considered, especially in contexts such as on smart devices and in public spaces where there isn't a screen to facilitate privacy policy engagement (Emami-Naeini et al, 2021; Helpful places, nd).

How meaningful consent will be realized as the use of AI expands complicates matters further, as AI developers are not always consumer-facing. This contributes to distance between data subjects and the organizations working with data sets, leading to the possibility of vast data sets used to train AI being built without ensuring the meaningful online consent of the data subject. These challenges may pervade AI development processes. As Crawford (2021: 95) notes, 'The AI industry has fostered a kind of ruthless pragmatism, with minimal context, caution, or consent-driven data practices while promoting the idea that the mass harvesting of data is necessary and justified for creating (AI) systems'. It is challenging enough to address these concerns when a single company collects data from users for its own purposes. More complex are examples such as Amazon's data exchange programme, where data sets are shared between data set providers and subscribers, creating additional distance between opportunities for consent protections and the future of AI development.

As policy makers consider the future of information protections, consent provisions must remain central to international law and regulation. The extent to which individuals realize forms of data power will be linked to whether forms of consent delivered are meaningful. Indeed, how to reduce 'the biggest lie on the internet' remains one of the central information policy challenges of the ongoing debate over the future of AI, and should be a primary focus of privacy research and advocacy.

Speaking back to corporate power: embedding opportunities for refusal – Ana Pop Stefanija

'What could I have done: Refused?' wrote one of my participants in her diary recording and detailing her interaction with a social media platform. Trying to figure out what data this platform holds about her, what kinds of inferences are made about her, and how her online and offline life is entangled with(in) the platform, she underwent a months-long process of going back and forth with the platform to obtain 'her' data. Becoming aware of her 'little to no leverage in this relation', she concludes: 'During these months of corresponding, I've gotten absolutely nowhere closer to obtain, what I wanted, but have remained absolutely where they wanted: as the powerful processors (owners) of MY data'.

And she is not the only one of my participants sharing this feeling of powerlessness. Navigating datafication and algorithmic systems and 'taking care' of one's data and one's entanglement with these systems increasingly felt like a futile and frustrating endeavour (see Pop Stefanija and Pierson (2023), for more detailed accounts). Anyone who wants to investigate this is faced with gatekeeping practices of platforms, with hidden 'entry points' to data about themself, with controlled flows of information even when access is 'authorized', with intentionally misleading, incomplete, incomprehensible, or overwhelming information, with obfuscated automated decision-making processes and absence of opportunities for inspection and redress.

These are just some of the rich insights collected through a participatory study that I conducted in 2020 with 47 participants, focusing on digital and social media platforms. Since I intentionally aimed towards a research design that enables a purposeful interaction and provides essential insights based on real-life experiences, participants were given diaries to capture their interactions, thoughts, and experiences while trying to figure out their position within and in relation to particular social media platforms. This setting, together with the adoption of the approach of critical companionship (Ziewitz and Singh, 2021), understood as a methodology for studying the lived experiences of individuals and as a research-with-care by providing support along the way, enabled a research design that was both about and with people.

The detailed diary reports and the time frame that enabled introspection and reflection of the participants, provided rich insights from which a number of concepts emerged. One element, the element of refusal, figured prominently. As envisioned by the participants and described in their diaries, this concept has two distinct arrangements: refusal as a practice and refusal as an opportunity.

Refusal as a practice relates to the wish or the need of the participants to be able to refuse. But refuse what? According to participants' diaries, this refers to being able to refuse having data collected about them (in general or by particular entities); to refuse the use or sharing of these data; to refuse to be profiled algorithmically; and, in general, to refuse to be subjected to algorithmic decision-making (see Pop Stefanija and Pierson, 2023).

The ability to refuse can be described as an opportunity for 'getting in the way', to borrow Ahmed's (2023) phrase. This strategy of getting in the way of datafication, profiling, and steering based on data and through algorithmic systems, means also having the ability to critically engage with one's data and the outputs of algorithmic systems. It also means an ability to make a decision for oneself, based on self-reflection and self-determination. It can also be seen as 'speaking back' – a process of being able to correct data inputs and algorithmic outputs and impose one's own version of 'truth' about oneself. This ability to 'speak back' is hence intricately related to the existence of opportunities to refuse.

These opportunities to refuse are related to refusal taking form – in order to be able to practise refusal, there must exist possibilities to refuse, these need to be afforded in the first place. This affording should be enabled via a number of elements and mechanisms (Davis, 2020). As the participants envision it, refusal should be located in the 'materiality of the medium' (Bucher and Helmond, 2018: 240), enabled primarily via the interface of the platform. As such it should take the form of tabs, buttons, pop-ups, visualizations, settings, and reminders, among others. These elements should enable the individuals to inspect, control, restrict, and opt-out of data collection practices, but also to modify, change, delete, and repair data inputs and algorithmic outputs. Some of these are already foreseen in the EU's GDPR (European Commission, 2016), however, their implementation in practice is lacking proper compliance (see, for example, noyb's open cases with EU's Data Protection Authorities (DPAs) (noyb, 2023)).

Being overpowered by corporate interests, agendas, and profit-making goals of private companies, individuals do not have much manoeuvring space around the datafication and algorithmic networks that influence their lives. For the moment, the acts of refusal and resistance, if possible at all, are experienced as labour intensive, requiring a lot of resources and time. They also require (almost expert) knowledge to navigate and understand the inner workings of algorithmic systems, as well as specific competences and particular capabilities and skills. These vary from initially having the knowledge that one has been algorithmically profiled, to knowing how to look and ask for one's data (for example, filing a Subject Access Request according to GDPR's Article 15, or using the platform's transparency tools), how to read the files (often times these are in unfamiliar formats, like JSON), to how to access these files (if possible, these are often hidden behind tabs and settings), and all the way to how to file and start a redress or complaint procedure (more information on these gatekeeping practices can be found in Pop Stefanija (2023)). Additionally, they sometimes require privilege to navigate the refusal process and to refuse at all (for example, see Jansen in this chapter). However, the ability to refuse for the participants of my study was intrinsically related to individual autonomy, control, self-reflection, self-directedness, and ultimately power. Power in relation to, or power over, the tech proprietors of these algorithmic systems, but also, power over oneself, as a power to act and steer one's actions and life in a self-determined manner. Designing algorithmic systems that embed opportunities for refusal should ensure that individuals will always have 'the chance to refuse' (Benjamin, 2016), if and when they want to.

Collective agency in the face of data + state power – Fieke Jansen

'Why is my son on that list?' (Peled, 2022) asks a mother whose son was selected for the Top400, a youth crime prevention approach of the city of Amsterdam in the Netherlands. From 2015 onwards, children and young adults who, in the eyes of the police and the city of Amsterdam, showed concerning behaviour, were selected for a crime prevention approach that combines care and control. Once selected, the municipality and its partners structurally intervene in their lives for a minimum of two years. This approach encroaches on those profiled and there are concerns with the way the Top400 criminalizes antisocial and teenage behaviour, instrumentalizes care for crime prevention, stigmatizes the youngster and their family, limits or obstructs access to justice and redress, and places a spotlight on their younger brothers and sisters (Jansen, 2022). The case of the Top400 offers insights into what a struggle for justice in the face of data + state power entails, as it is both a story of state repression and of resistance.

Between 2016 and 2019, over 300 children and young adults were selected for the Top400 through two data models, one of which was ProkidPlus.

This model identified a 'softer' group of 'at risk' youngsters, those that had been in contact with but not arrested nor charged by the police. Freedom of Information Act (FOIA) documents revealed that the inclusion of this 'softer' group of 125 Prokid children was hard to explain. The city advised civil servants responding to the question 'Why is my child selected for the Top400' not to mention the word algorithm or the name Prokid (Jansen, 2022). The municipality deliberately obfuscated the basis on which the 125 children were selected and limited their ability for redress. This case is not unique, it is just one of many stories where new forms of algorithmic governance (Dencik et al, 2019; Katzenbach and Ulbricht, 2019; Amoore, 2020) mediate and obfuscate decision-making in the European welfare state. However, the experiences of challenging the Top400 reveal that collective empowerment is a prerequisite for the struggle for justice in contemporary society.

Contemporary legal, technical, and social responses that aim to minimize the negative externalities of data power are connected to rights, knowledge, and skills of the individual. One prominent empowerment angle is that of increased data literacy, where building competencies will increase a person's ability to economically and socially participate in society (Pangrazio and Sefton-Green, 2020; Sander, 2020). Data literacy is conceptualized as more than learning how to read and write code or how to use new technologies, it is about a person's ability to navigate the complexities of contemporary societies. As such, most data literacy approaches aim to build knowledge and skills that will allow them to make informed choices about their digital lives. This approach assumes that with increased competencies people can directly control and influence the relationship between them and the data processor (Viljoen, 2020). That they just need to skill up their knowledge on data power and that they know how, and are able to, participate in political and social structures that enable and constrain datafication (Jansen, 2021). Reflections on our ongoing investigation into the Top400 reveal that data literacy is not enough in the struggle for justice in the face of data + state power.

In 2020 I met the documentary filmmaker Nirit Peled who had at that time spent over four years researching the Top400. She was struggling to tell its story. Public officials from the municipality and police did not want to go on the record, the mothers of the Top400 boys were afraid that speaking up publicly would lead to more stigmatization and reprisal, and the ex-Top400 boys wanted nothing to do with it, that time in their lives, or the state. Yet, by listening, Nirit Peled noticed that there was a discrepancy between the bureaucratic reality and success the city attributed to the Top400 and the lived experiences, foregrounding a number of serious concerns. To unravel the Top400, a collaboration was formed between the documentary filmmaker, a human rights lawyer, and me. We started a collective investigation into the Top400, where we requested and systematically analysed 2,000+ pages of

FOIA documents to gain more insights into the politics, the problems, the governance, and the data models behind the Top400. In November 2022 our investigation was made public through the documentary *Mothers* and the Top400 report (Jansen, 2022).

What we learned along the way was that Young's (2011) approach of listening to those subjected to state power to locate injustice(s) is the first crucial step in the long process of collective agency. It takes resilience, courage, and determination of those impacted to speak up against injustices caused by a paternalistic and repressive welfare state (Vonk, 2014). It took an interdisciplinary team years of research, conversations with those impacted, engaging with technical experts, and discussions with the wider network of stakeholders to make sense of the Top400. It took a documentary to be aired at a leading Dutch film festival and on national TV for the families to be heard and get some kind of recognition for the injustice of this intervention. It took the social and political capital of what society considers 'experts' to be able to put the problems of the Top400 on the political agenda. Despite the mayor's rejections of the injustice claims made by the community, the documentary, and the report (Halsema, 2022), the practice of investigation allowed for networks of solidarity with those impacted, the building of a collective, the foregrounding of injustices, and the articulation of justice claims.

The case of the Top400 shows that claiming justice in the face of state + data power is complex and cannot be the responsibility of informed individuals. When state power becomes enabled and enacted through data systems, the power asymmetry between the individual and the state increases. Challenging decisions of the datafied state requires knowledge and understanding of data systems, social capital, and political agency to claim rights, and resilience and courage to stand up. As such, I argue that we need to move away from the notion of individual empowerment through data literacy to collective agency through the practice of resistance. Collective agency, which should be understood as a process that brings together different competencies needed to identify and uncover the problem and jointly work towards a solution.

Datafication and community activism – Roderic Crooks and Lucy Pei

Like other forms of overt political work enacted in the register of the technological, data activism has been lauded as a vector of participatory power, including the power to subvert existing economies of knowledge production and expertise (Dencik et al, 2019; Lehtiniemi and Haapoja, 2020). But like other kinds of collective action, data activism arises from particular social locations, from people and communities dealing with the persistent consequences of structural inequality. As critical scholars have argued persuasively, the potentials of data to contribute to movements for justice (social or otherwise) are always tempered by competing, incommensurable understandings of what data can do, what kinds of political action are available to motivated parties in the present, and what kinds of people are considered legitimate civic participants: these dynamics favour those already privileged by economic and political hierarchies (Coleman, 2017; Gray, 2018; Heeks and Shekhar, 2019). In minoritized communities, however – those communities marked by socially consequential, interlocking forms of difference – building people-power via commercial tools and platforms of the tech sector poses specific risks. Minoritized communities, bound as much by material, intersecting differences such as race, class, gender, sex, citizenship, geography, and/or disability as by 'their ordered relation to capital' (Allen, 2021: 6), are the site of many forms of harm specific to the use of data-intensive computation. Chief among these perhaps is the datalogical enframing of community-defined problems, the brute mistranslation of the knowledge and experiences of working-class communities of colour into structures, documents, and evidence valorized by the state, academia, and the tech sector.

Since 2019, the Evoke Lab at UC Irvine has hosted an event called 'Datafication and Community Activism', a space where scholars and community organizers have been thinking about the relationship between minoritized communities and datafication. Community organizers work to shape voice and political strategy in the communities they serve. In the context of American political life, professional community organizers work in all kinds of communities in all parts of the political spectrum. The organizers we work with most often are based in minoritized communities, where they pursue larger social movement goals under precarious employment conditions in the not-for-profit sector. From the perspective of these organizers, to say a community is minoritized is to point to hierarchy in public life, to the way the public sphere is defined, constituted, and shaped for the benefit of dominant groups whose interests are enforced by the state, not by demography. This work has resulted (only infrequently) in meaningful and mutually beneficial working relationships with individual community organizers and community-based organizations of different kinds. Our work with organizers frequently puts us in a position to confront 'academic and nonprofit complicity' in the production and sale of data-intensive technologies that are used to harm minoritized communities via surveillance, criminalization, discrimination, and extraction.

Over the years, our strategy has prioritized listening to our colleagues so that we might be educated and informed about how digital data in all of their forms and manifestations relate to the self-determination and ongoing freedom struggles of working-class communities of colour. To date, we have been fortunate to work with many national, regional, and neighbourhood organizations who are generally interested in digital data but are specifically concerned with abolition of police, racial discrimination in computational systems, data-driven government service, reintegration for formerly incarcerated people, economic empowerment, and many other issues of interest to working-class communities of colour. These aspects of community organizing play out in very surprising ways, especially where the overtly political and social movement-aligned work of community organizing intersects with the pleasures and potentials of digital data. From the perspective of these organizers, datafication is not an unintended consequence of unpredictable technological change, but a continuation of the exploitation of working-class communities of colour. If data justice concerns 'fairness in the way people are made visible, represented and treated as a result of their production of digital data' (Taylor, 2017: 1), the community organizers with whom we work would remind technologists, academics, and civil society groups that digital technologies are inextricably linked to both state violence and private discrimination. Fairness, for many kinds of people in the United States, has never been on offer.

Participatory governance of datafication – Arne Hintz

The increasing use of data analytics for a variety of both commercial and public services occurs mostly without the knowledge of data subjects. We are profiled, categorized, assessed, sorted, and scored according to criteria that we do not understand, through processes that remain obscure, with consequences that are difficult to foresee, and with few possibilities to object or resist. This is already problematic in the context of commercial systems, such as the allocation of platform services and discriminatory pricing (Redden, 2022), but it becomes a fundamental challenge to democratic systems if it affects state functions and state–citizen relations. If our performance as citizens is permanently assessed through data systems, power is conferred onto the data collector (the state) and shifted away from citizens (Hintz et al, 2019). As a consequence, the role of the sovereign – the people – is diminished and citizens lose influence over government and public decision-making. This raises significant questions regarding people's roles in the deployment and management of data systems. How can, and should, we participate as citizens in governance systems that are informed and infused by data and AI? How do we intervene into decision-making about the roll-out of data and AI in government and the public sector? How do we advance civic agency and democracy in the datafied state?

Practices of, and research on, 'democratic innovations' offer a possible way forward in exploring how to advance citizen voices in decision-making outside and beyond established processes of institutionalized democratic engagement (Smith, 2009). Citizen assemblies, citizen juries, citizen summits, deliberative polling, distributed dialogues, and similar models and practices bring together a small selection of the population for deliberation on key issues that society is facing. Supported by expert input, a smaller or larger (from around 15 to potentially over 1,000) group of people, often recruited to represent wider society, meet for a few days and develop proposals or decisions on the issue they have debated. These methods have increasingly been applied to engage citizens in discussions on the use of data analytics in areas such as health, policing, and criminal justice. Think tanks, civil society organizations, policy institutions, regulators, and government departments have commissioned or organized such initiatives to understand people's views and seek guidance on policies and applications.

Research by the Data Justice Lab (Hintz et al, 2022) has explored the significant promises and challenges of these practices, focusing on the UK where they have enjoyed particular prominence recently, in part as a response to scandals and wider dissatisfaction with excessive data uses. We found that non-expert citizens were, in fact, able to discuss a complex topic such as data and AI with sufficient depth and to develop thorough outcomes and policy recommendations. Participants largely viewed the experience as positive and empowering, with the rare opportunity to both learn about the subject and make their voices heard. As deliberative exercises, however, these initiatives typically do not transfer decision-making power to citizens. They provide a platform for contributing voices and concerns but do not involve a substantial power shift. In some cases, they amount to little more than an opinion poll, which underlines concerns regarding their possible use for 'participation-' or 'engagement-washing', that is, the legitimation of decisions taken elsewhere. Yet many of them do have either direct policy impact or broader normative influences on decision-making.

Despite their titles, initiatives such as citizen juries and citizen summits are typically not self-organized at grassroots level, and organizers have significant leeway in framing discussions. Possibilities for participants to define the agenda and move the goalposts of the debate are often limited, and some organizers have steered deliberations (explicitly or implicitly) towards an acceptance of data uses and a recognition of its value. Further, the goal of representing a cross-section of society comes at the expense of considering impacts on, and experiences of, particularly affected communities. People from impoverished, racialized, and otherwise marginalized backgrounds are impacted by datafication in specific ways and their voices are crucial in properly assessing data uses, but they are not always incorporated.

As this brief snapshot demonstrates, these practices come with significant shortcomings. Building on theories of participation, we may categorize them as 'partial participation' (Pateman, 1970), at best, and as a form of 'tokenism' (Arnstein, 1969) that may lead to an advisory role in policy- and decision-making. They empower participants to learn, share their views, and (ideally) affect policy, but they do not amount to a sharing of decision-making power.

While they enhance people's voices in a debate dominated by commercial and governmental actors, their characterization as a 'people's practice' requires qualifications. However, they contribute to a growing composite of strategies for civic participation, together with community and grassroots initiatives, civil society campaigns, emerging institutional data governance structures (such as data trusts and data cooperatives), technical approaches towards algorithmic accountability, and other strategies as explored in this chapter. Together, these different models, practices and initiatives reflect a growing recognition that data subjects need to be involved in decisions about data, and that those who are affected by datafication should steer its future development and deployment.

Conclusion: Contextual research practices – Stine Lomborg and Anne Kaun

Research on people's practices in the face of data power, such as the examples described earlier in this chapter, testifies to the value of centring the people implicated by data power operations to understand the promise and perils of datafication and automated decision-making systems, and people's myriad ways of making their stakes in datafied societies. As we have seen, people contest and enact data power both in what may be considered small, individual acts of engagement, and in communal forms and participatory processes.

The use of data and automated decision-making systems in public services and private companies is often motivated with quantitative measures of efficiency and resource-savings. In the context of public welfare, it is also justified with reference to fairness and equal treatment based on the assumption that bias and noise will be minimized or eliminated altogether by reducing human intervention in the decision-making process. But the large number of failed pilot projects, along with several scandals across countries pertaining to systems that are actually put into use, suggest that ideals and reality do not always match all that well. As a first response, a number of high-level ethical guidelines for AI implementation and use have been developed and ratified at international, national, and organizational levels to shape the governance of data-driven systems. Arguably, guideline-based approaches to ethics and justice, too, assume standardization is not only possible for systems, but also for our ethical management and use of such systems. But the people who are actually implicated seem to be largely absent from these top-down and often universalizing discussions of the ethical uses of data, for example in relation to automated decision-making systems. This is problematic, because we know from decades of research about relationships between technology, data, and people that outcomes are not the same for everyone, and may vary substantially between social and cultural contexts. It remains to be seen whether and how systems that are built on clearly defined rules and enhanced standardization leave room for the different needs to reach capabilities for citizens.

Developing in-depth empirical accounts of people's data practices and experiences presupposes a move to radical contextualization. Context matters for how people make sense of data practices, what data practices they perceive as fair and just, and what they are capable of doing with or to them. Context also matters for whether people experience a controlling state or commercial actors as their main combatant. This chapter, therefore, ends with putting forward a programmatic statement for the need for radical contextualization as a way of centring people, and giving voice to people in all their diversity in discussions of data power. In terms of a research agenda this means 'to consider the specific contexts within which people (professionals as well as private citizens) interface with ADM [and other data-based technologies], the ways in which they make sense or reject them and ultimately develop frameworks for approaching technological continuity and change' (Lomborg et al, 2023: 14).

Attending empirically to the lived experiences and stories of agentic, ambivalent, and alienated human beings, as demonstrated across this chapter, can help us better understand the social and cultural-contextual dynamics of data power and data-induced empowerment. And it can amplify people's voices in public debates on data power, which might lessen the burden on individuals when facing data power. In turn, however, it also demands reflection on researcher roles, and the nexus between science, community engagement, and activism when working with people front and centre.

Commitment to contextualization entails a shift in what is figure and what is ground as we study data power. How can we centre implicated people and their positions while keeping an analytical eye on the data infrastructures that they act upon? How can we do justice to cultural contingencies, social exigency, and historical trajectories that shape people's practices in the face of data power, without compromising culturally comparative scholarship? People-centred approaches must balance these needs to bring lived experience in dialogue with studies of the platforms and actors who channel, retain, augment, and consolidate data power in pursuit of competitive edge in innovation, population management, and economic gain.

A people-centred and radically contextualized approach to data power will also allow to tease out the possibilities to agency in the data practices discussed earlier – consent practices, refusal practices, citizen participation (including citizen juries and citizen assemblies), as well as other forms of data activism. Such an approach will allow to disentangle structural inequalities in the current data ecologies while highlighting possibilities of agencies and shifts from the bottom up. By extension, this underlines that there is nothing natural about the state of our data-based systems, rather that they are always changeable and in the making even if such changes require immense engagement and work.

Discussant – Catherine D'Ignazio

The strategies outlined in this chapter are exciting, both because they represent diverse forms and scales of resistance to data power and also because they come from empirical and participatory work with people and communities. As Aristea Fotopoulou has argued, researchers should look to shift their object of research from data, algorithms, and platforms themselves towards the human practices of acquiring, analysing and using data, so that we may 'reinstate the materiality of data, to think about laboring bodies, invisible human practices, and social relations and activities' (Fotopoulou, 2019).

One issue that I would like to raise in relation to data activism relates to the diverse data epistemologies employed by activists themselves. In this chapter, the authors have mainly centred on data as it is acquired, stored, analysed, and deployed in models, algorithms, and predictions by state and corporate actors. The epistemological approach to data science embodied by states and corporations is just one approach – and it is one that is predominantly positivist, optimizing, and neoliberal. It results in, as Crooks and Pei state, 'surveillance, criminalization, discrimination, and extraction' (this chapter). To their list, I would add 'grotesque accumulation' because ultimately this approach is about wealth hoarding for those on top and the operationalization of scarcity and inequality for everyone else.

But as prior work has shown, this is not the only epistemology of data. In their work on data activism, Stefania Milan and Lonneke van der Velden (2016) call attention to the important role of data activists who function as 'producers of counter-expertise and alternative epistemologies, making sense of data as a way of knowing the world and turning it into a point of intervention. They challenge and change the mainstream politics of knowledge' (Milan and van der Velden, 2016: 63–64). In recent years, alternative epistemological approaches to data have been flourishing: data feminism, indigenous data sovereignty, Data for Black Lives, environmental data justice, QuantCrit, and queer data, to name a few. These approaches posit that there are other ways of using (or refusing) data in the service of co-liberation.

For the past four years, I have been working on a participatory research and design project called Data Against Feminicide in which my colleagues and I have been working with feminicide data activists in the Americas. These are groups – of academics, journalists, activists, nonprofits, concerned individuals, mothers, sisters, aunties, and families – who painstakingly document cases of fatal gender-related violence and use those data for a variety of political demands and impacts. The information ecosystem in which they work is deeply biased: states publish little to no information about feminicide, even where laws do exist. The state regularly misclassifies the killings of gender and racial minorities as accidents and suicides. The media – which often end up being activists' main sources of information – are racist, misogynist, transphobic, and victim-blaming.

And yet, activists persist in assembling carefully curated spreadsheets and databases. They seek humanizing photos of killed women. Some groups provide direct services and accompaniment through the justice system to families. Other groups stage collective memorials with empty chairs or empty shoes, using aggregated absence as an aesthetic approach. Still others use their data to gain audience with the state or to influence and reframe toxic media narratives about this violence.

As I have reflected on this work with my collaborators and the activists themselves, I have written about just how profoundly the activists' epistemological approach to working with data diverges from the mainstream positivist approach (D'Ignazio, 2024). Feminicide data activists centre care, memory, and justice. They are often from the communities from which they draw their data – they are women, Black women, indigenous women, Latin American women, trans women, survivors, mothers, family members, community members. Rather than using data to 'solve' a problem, they use data to remake and reframe the problem of feminicide. They challenge the idea that gender-based violence is a personal problem and they reframe it as a political problem, a public problem, a structural problem. Feminicide data activists are deeply aware of the biases and limitations of their data because they are, themselves, the data producers. They talk about their role as caring for the data and the people and lives represented therein.

Does producing carefully documented databases of feminicide cases 'solve' feminicide? No, and no activist would imagine that it does. It is an imperfect informatic tactic in a deeply asymmetrical environment. But this production of information does participate in a broader constellation of efforts that are working towards the restoration of rights, the healing of communities, and the longer-term work of structural transformation. One thing emphasized to me over and over again by activists is that this work is not about counting the dead, it is a defence of life itself.

I offer these thoughts on data epistemologies so that we do not forget to think differently about our information and our technologies. As with all tools, their origins do not preclude their appropriation, their reclamation, and their reimagination. As Paola Ricaurte frames it, there are 'possible alternative data frameworks and epistemologies that are respectful of populations, cultural diversity, and environments' (Ricaurte, 2019). These may help us work towards data practices in the service of life, living, and vitality.

Discussant – Dan McQuillan

By assembling and analysing a vivid canvas of people's actual practices, this chapter makes a valuable contribution to a critique of data power. I have tried to respond diffractively; that is, through a constructive approach to the differences present within the text and between the text and my own perspectives. For brevity this is presented as a series of statements; some of which, I hope, may resonate with the reader.

- It is clear from the chapter that datafication is an attack on the poor and the marginalized.
- Datafication renders social relations as abstractions for distanced and indifferent manipulation.
- Datafication is not a shift of power relations away from an acceptable norm but an intensification of existing injustices.
- Any claims that data are being collected to fulfil people's needs are a distraction and diversion.
- Datafication is degradation.
- We do not need a stake in the datafied society, we need to consciously and explicitly resist it.
- The extractive and centralizing logics described in the chapter reduce the space available for a livable life.
- Optimization and efficiency replace relationality with resource extraction.
- More data under datafication means more precarity and austerity.
- Datafication is part of material structures that burn through energy, create emissions, and deplete water resources, all while claiming to be a solution to the climate crisis.
- Calling for resistance is to recognize what is at stake; we resist in order to exist.
- A resistant framing includes people's existing practices but points beyond them.
- It is not an attempt to impose a programme but to challenge the datafied foreclosure of the future.
- Subjects in a datafied society are shaped by data power; practices of resistance restore different subjectivities.
- Resistant subjects do not simply have viewpoints but standpoints; valid forms of knowing that are embodied and situated, not abstract and distant.
- Where datafication drives precaritization, resistance responds with mutual aid.
- Where datafication leads to exclusion and oppression, resistance develops solidarity.
- Resistance is also a call for academic commitment rather than complicity; as the chapter demonstrates, research can remain rigorous while asking questions that are resistant.

- Resistance is more than collective refusal; it is a commitment to possible alternatives.
- Resistance is 'one no, many yeses'.
- Resistant responses to data power are those that develop counter-power.
- Consent becomes meaningful under conditions that Illich (1973/1975) called conviviality; where it enables autonomous action by means of tools least controlled by others.
- Anything other than conviviality merits collective refusal.
- Resistant refusal develops not only data literacy but self-organization literacy, critical feminist literacy, decolonial literacy, and so on.
- Resistance to datafication is a social movement, or rather, a movement that is part of other social movements where injustices are becoming datafied.
- Forms of 'consultation' and 'participation' that do not transform power relations in favour of these movements are actually modes of assimilation.
- Datafication is a slippery opponent because it facilitates fake empowerment, with proposals to participate in our own datafication and 'control' our own data in the name of datafication-for-good.
- As Ruha Benjamin (2019) says: 'Demanding more data on subjects that we already know much about is ... a perversion of knowledge in which the hunt for more and more data is a barrier for acting on what we already know.'
- Datafication is the new 'Society of the Spectacle', but its very pervasiveness creates opportunities for intersectional resistance.
- Where so many are affected in so many dimensions, resistance means making connections across different contexts of datafication.
- This resistance can learn from previous movements against repressive technologies, such as the Luddites (see, for example, Binfield, 2004), who were not anti-technology but anti-automatization, and who defended their autonomy through militant community mobilization.
- Resisting datafication is not about rejecting technology, but a matter of resolving for ourselves which technologies and which subjectivities will emerge together in response to which material problems.
- In other words, resistance to datafication is the development of a prefigurative technopolitics.
- Resistance to datafication is about responding differently to the problems that are currently being datafied, in ways that are both collective and technical.
- Whereas the constant demand for data always implies an existing deficit, not only of knowledge but also of capacity, resistance to datafication builds on our collective strengths and resilience.
- Like the workers who developed the Lucas Plan, it asks: what do we already know, what can we already do that can become transformative?
- Resistance to datafication is this search for new socialities and new tools, for practices that reclaim the common good.

- Datafication is rooted in fossil fuel modernity and the reduction of all beings to a standing reserve.
- Our resistance is not only the careful contextualization of existing dystopias, but the multiplication of alternative futures.

Notes

1. Alphabetical author order
2. Discussant
3. Facilitator
4. Discussant

References

Acquisti, A., Adjerid, I., Balebako, R., Brandimarte, L., Cranor, L. F., Komanduri, S., et al (2017). Nudges for privacy and security: Understanding and assisting users' choices online. *ACM Computing Surveys (CSUR)*, 50(3): 1–41.

Ahmed, S. (2023). *The Feminist Killjoy Handbook*. Allen Lane. Retrieved from: www.penguin.co.uk/books/454793/the-feminist-killjoy-handbook-by-ahmed-sara/9780241619537

Allen, J. S. (2021). *There's a Disco Ball between Us: A theory of black gay life*. Duke University Press.

Amoore, L. (2020). *Cloud Ethics: Algorithms and the attributes of ourselves and others*. Duke University Press.

Arnstein, S. R. (1969). A ladder of citizen participation. *Journal of the American Institute of Planners*, 35(4): 216–224.

Benjamin, R. (2016). Informed refusal: Toward a justice-based bioethics. *Science, Technology, and Human Values*, 41(6): 967–990.

Benjamin, R. (2019). *Race After Technology: Abolitionist tools for the New Jim Code*. Polity.

Binfield, K. (ed) (2004). *Writings of the Luddites*. Johns Hopkins University Press.

Bucher, T. and Helmond, A. (2018). 'The Affordances of Social Media Platforms.' In: J. Burgess, T. Poell, and A. Marwick (eds) *The SAGE Handbook of Social Media*, pp 233–253. SAGE.

Cate, F. H. (2006). The failure of fair information practice principles: Consumer protection in the age of the information economy. Retrieved from: https://papers.ssrn.com/sol3/papers.cfm?abstract_id=1156972

Cohen, J. E. (2019). *Between Truth and Power: The legal constructions of informational capitalism*. Oxford University Press.

Coleman, G. (2017). From internet farming to weapons of the Geek. *Current Anthropology*, 58(Supplement 15): S91–S102.

Crawford, K. (2021). *Atlas of AI*. Yale University Press.

D'Ignazio, C. (2024). *Counting Feminicide: Data feminism in action*. MIT Press.

Davis, J. L. (2020). *How Artifacts Afford: The power and politics of everyday things*. MIT Press.

Dencik, L., Hintz, A., Redden, J., and Treré, E. (2019). Exploring data justice: Conceptions, applications and directions. *Information, Communication & Society*, 22(7): 873–881.

Draper, N. A. and Turow, J. (2019). The corporate cultivation of digital resignation. *New Media & Society*, 21(8): 1824–1839.

Emami-Naeini, P., Dheenadhayalan, J., Agarwal, Y., and Cranor, L. F. (2021). An informative security and privacy 'nutrition' label for internet of things devices. *IEEE Security & Privacy*, 20(2): 31–39.

Eubanks, V. (2017). *Automating Inequality: How high-tech tools profile, police, and punish the poor*. St Martin's Press.

European Commission (2016). Regulation (EU) 2016/679 of the European Parliament and of the Council of 27 April 2016 on the protection of natural persons with regard to the processing of personal data and on the free movement of such data, and repealing Directive 95/46/EC (General Data Protection Regulation). Retrieved from: https://eur-lex.europa.eu/eli/reg/2016/679/oj

Fotopoulou, A. (2019). 'Understanding Citizen Data Practices from a Feminist Perspective: Embodiment and the Ethics of Care.' In: H. Stephansen and E. Trere (eds) *Citizen Media and Practice: Currents, connections, challenges*, pp 227–242. Taylor & Francis/Routledge.

Gray, J. (2018). Three aspects of data worlds. *Krisis: Journal for Contemporary Philosophy*, 1: 4–17.

Habib, H., Pearman, S., Wang, J., Zou, Y., Acquisti, A., Cranor, L. F., et al (2020). 'It's a scavenger hunt': Usability of websites' opt-out and data deletion choices. *Proceedings of the 2020 CHI Conference on Human Factors in Computing Systems*, pp 1–12.

Halsema, F. (2022). Raadsinformatiebrief: Recente media-aandacht voor de Top400-aanpak. Gemeente Amsterdam. Retrieved from: https://open.overheid.nl/documenten/ronl-8f2884ba05c31a13f2a7e463932502d81289fde7/pdf

Hargittai, E. and Marwick, A. (2016). 'What can I really do?' Explaining the privacy paradox with online apathy. *International Journal of Communication*, 10: 3737–3757.

Heeks, R. and Shekhar, S. (2019). Datafication, development and marginalised urban communities: An applied data justice framework. *Information, Communication & Society*, 22(7): 992–1011.

Helpful Places (nd). Digital Trust for Places & Routines. Retrieved from: https://dtpr.io

Hintz, A., Dencik, L., and Wahl-Jorgensen, K. (2019). *Digital Citizenship in a Datafied Society*. Polity Press.

Hintz, A., Dencik, L., Redden, J., Trere, E., Brand, J., and Warne, H. (2022). Civic Participation in the Datafied Society: Towards Democratic Auditing? Research Report. Retrieved from: https://datajusticelab.org/wp-content/uploads/2022/08/CivicParticipation_DataJusticeLab_Report2022.pdf.

Illich, I. (1973/1975). *Tools for Conviviality*. Fontana.

Jansen, F. (2021, August). Critical is not political: The need to (re) politicize data literacy. Seminar.net, 17(2).

Jansen, F. (2022). Top400: A top-down crime prevention strategy in Amsterdam. PILP. Retrieved from: https://pilpnjcm.nl/wp-content/uploads/2022/11/Top400_topdown-crime-prevention-Amsterdam_v2.pdf

Katzenbach, C. and Ulbricht, L. (2019). Algorithmic governance. *Internet Policy Review*, 8(4): 1–18.

Kennedy, H. and Moss, G. (2015). Known or knowing publics? Social media data mining and the question of public agency. *Big Data & Society*, 2(2): 1–11.

Lannerö, P. (2012, 27 January). Previewing online terms and conditions: CommonTerms alpha proposal. Retrieved from: http://commonterms.org/commonterms_alpha_proposal.pdf

Lehtiniemi, T. and Haapoja, J. (2020). Data agency at stake: MyData activism and alternative frames of equal participation. *New Media & Society*, 22(1): 87–104.

Lomborg, S., Kaun, A., and Hansen, S. S. (2023). Automated decision-making research: Towards a people-centered approach. *Sociology Compass*, 17(8): e13097. https://doi.org/10.1111/soc4.13097

McDonald, A. M. and Cranor, L. F. (2008). The cost of reading privacy policies. I/S: *A Journal of Law and Policy for the Information Society*, 4(3): 543–568.

Milan, S. and van der Velden, L. (2016). The alternative epistemologies of data activism. *Digital Culture & Society*, 2(2): 57–74.

noyb. (2023, 21 July). Projects | noyb.eu. Noyb – None of Your Business. Retrieved from: https://noyb.eu/en/projects

Obar, J. A. (2022a). A policy complexity analysis for 70 digital services. Retrieved from: www.biggestlieonline.com/policy-complexity-analysis-2019/

Obar, J. A. (2022b). Unpacking 'the biggest lie on the internet': Assessing the length of terms of service and privacy policies for 70 digital services. Retrieved from: https://papers.ssrn.com/sol3/papers.cfm?abstract_id=4293363

Obar, J. A. (2023, March). Consumer Privacy Protection Act could lead to fines for deceptive designs in apps and websites. *The Conversation*. Retrieved from: https://theconversation.com/consumer-privacy-protection-act-could-lead-to-fines-for-deceptive-designs-in-apps-and-websites-196019

Obar, J. A. and Oeldorf-Hirsch, A. (2018). The clickwrap: A political economic mechanism for manufacturing consent on social media. *Social Media + Society*, 4(3): 1–14.

Obar, J. A. and Oeldorf-Hirsch, A. (2020). The biggest lie on the internet: Ignoring the privacy policies and terms of service policies of social networking services. *Information, Communication & Society*, 23(1): 128–147.

Obar, J. A. and Oeldorf-Hirsch, A. (2022). Older adults and 'the biggest lie on the internet': From ignoring social media policies to the privacy paradox. *International Journal of Communication*, 16: 4779–4800.

OPC (Office of the Privacy Commissioner of Canada) (2020, August). PIPEDA fair information principle 3 – consent. Retrieved from: www. priv.gc.ca/en/privacy-topics/privacy-laws-in-canada/the-personal-info rmation-protection-and-electronic-documents-act-pipeda/p_principle/ principles/p_consent/

OPC (Office of the Privacy Commissioner of Canada) (2021, August). Guidelines for obtaining meaningful consent. Retrieved from: www.priv. gc.ca/en/privacy-topics/collecting-personal-information/consent/gl_ omc_201805/

Pangrazio, L. and Sefton-Green, J. (2020). The social utility of 'data literacy'. *Learning, Media and Technology*, 45(2): 208–220.

Pateman, C. (1970). *Participation and Democratic Theory*. Cambridge University Press.

Peled, N. (2022). 'Waarom staat mijn zoon op die lijst?' Groene Amsterdammer. Retrieved from: www.groene.nl/artikel/waarom-staat-mijn-zoon-op-die-lijst

Pop Stefanija, A. (2023). 'Power Asymmetries, Epistemic Imbalances and Barriers to Knowledge: The (im)possibility of knowing algorithms.' In: S. Lindgren (ed) *Handbook of Critical Studies of Artificial Intelligence*, pp 563–672. Edward Elgar Publishing.

Pop Stefanija, A. and Pierson, J. (2023). Algorithmic governmentality, digital sovereignty, and agency affordances: Extending the possible fields of action. *Weizenbaum Journal of the Digital Society*, 3(2): 1–30. https://doi. org/10.34669/WI.WJDS/3.2.2

Redden, J. (2022). 'Data Harms.' In: Dencik, L., Hintz, A., Redden, J., and Trere, E., *Data Justice*, pp 59–72. SAGE.

Reidenberg, J. R., Breaux, T., Cranor, L. F., French, B., Grannis, A., Graves, J. T., et al (2015). Disagreeable privacy policies: Mismatches between meaning and users' understanding. *Berkeley Technology Law Journal*, 30(1): 39–68.

Ricaurte, P. (2019). Data epistemologies, the coloniality of power, and resistance. *Television & New Media*, 20(4): 350–365.

Roosen, M. (2020). What SyRI can teach us about technical solutions for societal challenges. Global Data Justice, 20 February. Available at: http://globaldatajustice.org/2020-02-20-roosen-syri/

Sander, I. (2020). What is critical big data literacy and how can it be implemented? *Internet Policy Review*, 9(2).

Smith, G. (2009). *Democratic Innovations: Designing institutions for citizen participation.* Cambridge University Press.

Srnicek, N. (2017). *Platform Capitalism.* John Wiley & Sons.

Taylor, L. (2017). What is data justice? The case for connecting digital rights and freedoms globally. *Big Data & Society*, 4(2): 1–14.

TOS;DR (Terms of Service; Didn't Read) (2023). Frontpage. Retrieved from: https://tosdr.org

Viljoen, S. (2020). Democratic data: A relational theory for data governance (SSRN Scholarly Paper ID 3727562). Social Science Research Network. https://doi.org/10.2139/ssrn.3727562

Vonk, G. (2014). Repressive welfare states: The spiral of obligations and sanctions in social security. *European Journal of Social Security*, 16(3): 188–203.

Young, I. M. (2011). *Justice and the Politics of Difference.* Princeton University Press.

Ziewitz, M. and Singh, R. (2021). Critical companionship: Some sensibilities for studying the lived experience of data subjects. *Big Data & Society*, 8(2). https://doi.org/10.1177/20539517211061122

5

Practitioner Interventions in Data Power

*Hadley Beresford,[1] Iris Muis,[2] Susan Oman,[3] Joanna Redden,[4]
Elise Renkema,[5] Marlee Tichenor,[6] Caitlin Bentley,[7]
Teresa Cerratto-Pargman,[8] and Jo Bates[9]*

Introduction

Are we entering a new phase of urgency over crises in ethical data practices? Indeed, what is data work? At the time of writing, the media is filled with predictions and proposals that generative artificial intelligence (AI) will be responsible for the end of the world. More mundanely, there has been much concern that increasingly automated processes will displace both unskilled and skilled labour. This includes a variety of data work. This chapter reflects on the nature of data work: what it looks like in practice and how it is differentiated across various data practitioners. We do so from a critical data studies position and the hope that in reflecting on the growing variety of practitioners, we may contribute to the data work that intervenes in data power.

Data work involves an ecosystem of interdependent roles, professions, and actors, as Beresford illustrates below. They draw from across domains and sectors, explaining that while policy makers and others in power are increasingly called upon to reflect on what good or fair data practices may involve, data scientists are increasingly called upon to create solutions that fit whatever neat or narrow criteria may be set out. Tichenor's research with those who identify as part of a 'global statistical community'[10] found that, while this group would not identify as data scientists as a rule, their professional identity was dependent on ideas that they were doing good with data. Yet, they were resistant to the rise of data science, as something at odds with traditional statistics as a practice of governance that has used data for hundreds of years, and driven by an ethics of care for those found within the data. These participants were reflective about the negative impacts of their work on certain populations. In short, they felt accountable for their counting practices and acted with a generative tension with various framings of data work.

Research with council members in the Netherlands, by contrast, found that these people who work with data to make decisions that affect populations distanced themselves again. Renkema and Muis suggest that these people who work with data consider themselves 'laypersons', who are indeed not data professionals, thus becoming unaccountable for decisions made. When is someone data literate 'enough' to make decisions about data and algorithms? And, who decides what the demarcation of data literacy is in local governance? Ask Renkema and Muis.

Oman comments on the UK government's line that they were 'following the data' to deny accountability when making controversial policy decisions, particularly during the COVID-19 pandemic. The empirical research she discusses here, however, talks about those in the data ecosystem who have discomfort with their day-to-day data responsibilities, feeling that they do not have the data expertise to collect or process these data. In short, instead of worrying about data for good, or what a good data practice is, they worry about whether they are good enough to work with data.

The chapter concludes with a reflection on what data work is, and how training data scientists in being more reflective and aware of the impacts of the work of data may improve future data practices. Redden extends 'data work' to those important activities that activists, journalists, and members of affected communities have taken on in order to limit the negative effects of automated decision systems (ADS). Redden and her colleagues' work has shown that an often critical factor in the removal of a harmful ADS has been civil society critique and mobilization. In this way, we also assert that the work of critical data studies is data work in itself – as an attempt to capture what is hidden, nefarious, and/or negligently injurious in current practices with the hope that data futures can be better for all. Further, that critical data studies has much to gain by recognizing and learning with those engaged in political and power struggles about if and how data should be used in practice.

Data scientists and algorithmic bias mitigation in the public sector – Hadley Beresford

In recent years, the UK civil service, as well as the public sector more broadly, have moved towards further utilizing data-based approaches, specifically in the form of algorithmic technologies (Algorithm Watch, 2019; Big Brother Watch, 2021). The increased use of algorithmic technologies in the UK context has been at least somewhat influenced by the legacy of austerity policies – local councils and public-sector departments are having to 'do more with less' (Dencik et al, 2018; Oswald et al, 2018). Indeed, Eubanks (2018) notes that the increase of algorithmic technologies in public services has arrived alongside the rapidly rising economic insecurity of the last decade. As these technologies enter more mainstream use, cases of algorithmic bias have become more prevalent (Angwin et al, 2016; O'Neil, 2017; Eubanks, 2018). Algorithmic bias describes how, through a combination of social, technical, and probabilistic mechanisms, some people are penalized, or denied opportunities, due to their membership of a marginalized group.

Thus far, many of the solutions data scientists have produced to mitigate the risks of algorithmic bias have been technical or procedural in nature, including debiasing techniques which focus on ensuring datasets are either more representative of their target population, or relying on achieving statistical parity when comparing the outcomes of different groups based on protected characteristics (Galhotra et al, 2017). Other approaches have included finding new ways to operationalize the concept of fairness within a statistical framework, allowing data scientists to better perform statistical checks on their models (Bellamy et al, 2018).

These types of technical checks fit nicely into the skillsets of data scientists, however, these methods have been criticized by academics and third-sector organizations for not addressing the social structures and processes that contribute to algorithmic bias. Focusing on biases which may exist within datasets and statistical methods, critics argue, means inadequate attention is paid to the ways these systems embed biases from the socio-economic system the algorithm is deployed within (Balayn and Gürses, 2021). In Hoffmann's paper 'Where fairness fails', she critiques these types of technical solutions as positioning the problem as one which is caused by 'bad algorithms' which can be fixed, distracting from consideration of the wider socio-economic influences and logics these algorithms are embedded within (Hoffmann, 2019). Hoffmann, instead, calls for a focus on justice, rather than fairness, to consider these wider socio-economic influences (Hoffmann, 2019). Similarly, Green calls for data scientists to become cognizant of the political assumptions embedded within these technologies and question them within their own working practices (Green, 2021).

While these academic perspectives help distinguish between different ways of operationalizing the 'social good', and draw attention to implications attached to different framings, there is still uncertainty around the types of strategies public-sector data practitioners can utilize in their day-to-day work. When I have spoken to public-sector data scientists, many genuinely want to find working practices which produce 'good' and non-discriminatory algorithms. However, the definition of 'social good' promoted in these cultures is often heavily structured by the political and legal structures these data practitioners work within. Often, these practitioners are creating digital 'products' designed to serve a wide-ranging spectrum of citizens – each of whom will have their own conception of the 'social good'. Additionally, data practitioners find the types of approaches available to them are strongly influenced by legal structures (Orr and Davis, 2020). Furthermore, data practitioners are, unfortunately, having to 'do more with less', due to increasing workloads with little extra funding from government bodies (Oswald et al, 2018). These combined issues mean socially motivated practitioners find themselves trying to work towards some form of 'social good' amenable to the context they work within, with the outcome of this process often uncertain.

Moreover, it is not only the framing of these problems which restricts data practitioners' ability to work towards making tangible improvements in this area. Algorithmic technologies are not the providence of data scientists alone – these technologies are created using multidisciplinary teams, including policy experts, ethics experts, data engineers, administrative staff, and front-line staff responsible for data collection. Depending on the service, department, or project in question, these actors can be incredibly varied. This means effort is required to create a shared understanding of these things across stakeholders with often disparate skillsets, expertise, knowledge-bases, values, and beliefs. There is a need for practitioners to create interdisciplinary processes which allow them to address contextual organizational and structural constraints in public-sector algorithmic bias mitigation.

Sustainable development data practitioners – Marlee Tichenor

In many ways, in our data-laden world, we are all data practitioners, and as such we are always negotiating the ways that constantly collected and processed data help constitute our understandings of phenomena, both global and intimate. Thus, perhaps it is unsurprising that there would be such tension around who defines themselves a 'data practitioner' in the context of these phenomena constituted by data. However, to give it some specificity, the data practitioners with whom I most recently worked were interested in representing social, economic, and environmental problems as global problems, with the goal of intervening upon them and monitoring progress on these interventions. They were trained in statistics, international development, or economics, and they called themselves members of the 'global statistical community', connected by the annual United Nations Statistical Commission (UNSC) and its various working groups, and currently shaped by the context of the Sustainable Development Goals (SDGs).

The SDGs were shaped and agreed upon by the United Nations General Assembly, whose deliberations were fed in by various consultancy processes. These processes included the UN Secretary General-run 'Post-2015

Development Agenda' – whose goal was to build a next-generation version of the Millennium Development Goals (MDGs), and which has been widely criticized as a donor-driven set of goals largely directed at low- and lower middle-income countries (for example, Amin, 2006; Saith, 2006) – and the Open Working Group that emerged from the Rio+20 Conference on the Environment and Development – which was driven by a country-led coalition and whose goal was to foment more structural change (Fukuda-Parr and McNeill, 2019). However, it was also driven by an increased attention to official statistics and development data in the global governance space – partly fuelled by the previously mentioned MDGs and the monitoring of their indicators towards success – a flurry of initiatives and declarations whose apex might be best captured in the UN's report, 'A world that counts: Mobilising the data revolution for sustainable development' (IEAG, 2014). In this way, at least rhetorically, the SDGs were to be shaped by the twin revolutions of data-driven development and of country-led coalitions from the Global South to reverse the usual top-down approach to development (Bandola-Gill et al, 2022).

The sustainable development data practitioners driving that first engine made up a heterogeneous group, and most of them would not consider themselves data scientists. In fact, in my interviews with them, it was frequently mentioned that there was palpable tension and discomfort in the community about the rise of data science in the last 15 years, in this space that has been shaped by statistical thinking since the first annual UNSC in 1947 – what one interlocutor called "the traditional official statistical universe" (UN Statistician, 3).[11] As my colleagues and I have shown elsewhere, these data practitioners and the work that they do are shaped by three kinds of reflexivity: epistemic, care-ful, and instrumental (Bandola-Gill et al, 2023). Many of our interlocutors were quite aware of the performative nature of the decisions they made about the global measurement of social, economic, and environmental problems (Callon, 2006) – they acknowledged the epistemic limits of measurement and the effects of these decisions on 'distorting' policy on the domestic level down the line. We found that, for many of them, their work was also framed by a 'duty of care' to those whose lives they represented in numbers, precisely to attempt to minimize the harm produced by the limits of numbers – they were motivated by caring for these populations through number production. Finally, we found that these two forms of reflexivity became resources for these actors to use to negotiate the politics around promoting particular agendas – they instrumentalized reflexivity as a way to build trust with stakeholders and to build consensus.

While engaging in these reflexive data practices, tensions arose between the intention of caring for these populations through enumeration and data disaggregation, and the outcomes of these practices of enumeration and data disaggregation. This includes the frequent conceptual collapsing of the practice of enumeration with addressing social harms, a framing that is often used to advocate and seek funding for the labour and materials needed for constant monitoring and enumeration. As one interlocutor acknowledged, with the case of building and supporting civil registration and vital statistics (CRVS) systems globally – systems which are less robust in many parts of the world – there is this perception produced by "those [CRVS] guys" that "if one got to universal civil registration and timely production of vital statistics, magically … it'll just have a flow on effect and so many other development problems and health problems will be resolved" (UN Statistician, 13). He noted that this illusion of collapsing measurement with solving development problems was a "huge Achilles' heel" of efforts to promote effective statistical capacity around the SDGs and global agendas in general. In this way, these sustainable development data practitioners conducted their work through tensions in their identities as such, as well as in the gaps between their actions' efforts and their felt effects.

Democratic control over data and AI projects in the local public sector – Elise Renkema and Iris Muis

In this chapter, we share our initial findings and experiences with interventions in the field of datafication and algorithmization of policy making and the role of elected representatives within this trend. We argue for an empirically driven and socially engaged practice of critical data studies, covering local contexts of data practices. This provides in-depth insight into the discourses of data power, facilitating effective knowledge transfer and social engagement for building a fair and open digital society.

During our work with municipalities in the Netherlands, we focused specifically on elected representatives who were seated in city councils. We interacted with them through teaching masterclasses around themes of datafication, algorithmization, and democratic control. An example of this is the masterclass series for city council members of the city of Utrecht, which consisted of four two-hour meetings. An array of topics was discussed, from the impact of sharing platforms on public space to formulating a political stance on prediction algorithms.

Most research in the field of critical data studies in government focuses on policy officers (civil servants), but not a lot of work has been done on researching the role of elected representatives in the datafied society. There has only been one significant research report on the role of city council members (Das et al, 2020), its main conclusion being that they lacked sufficient knowledge and expertise to fulfil their role as elected representatives when it comes to processes of digital technology. This is a conclusion we also made during our research.

We have identified several challenges of council members to fulfil their representing and controlling role. One main issue is how council members perceive their relationship to data and algorithms. Council members often say they are fulfilling their role in the council as 'laypersons', someone who is not a professional in the given field. We have noticed that council members are often either unsure of their ability to make policy or control decisions about data and algorithms or do not see that it is their responsibility. This brings us to the question: when is someone data literate 'enough' to make decisions about data and algorithms? And, who decides what the demarcation of data literacy is in local governance? What complicates this matter is that the political arena is often not considered a 'safe space' to learn or make mistakes.

A second issue is that not all council members see that data and algorithms belong in the political arena, due to it being seen as a matter of execution and a non-political issue and therefore not their responsibility. "Leave the IT questions to the IT experts, I do not even know how to properly operate my own laptop", is something we often hear. We have also noticed that some council members perceive that data can be seen as a neutral tool. As a result, they have not developed an ideological vision on this matter. Moreover, the issue might not reach the political agenda of the council because it stays in the administrative part of local government.

Overall, this can be seen as a question of identification. Do elected representatives see themselves as competent enough to provide democratic control over data projects and algorithms? And, are such projects identified as part of the political arena by elected representatives? We argue for more research on how the domain of digitalization and algorithmization can be embedded in existing democratic institutional processes. Such research should be embedded in the local contexts of data practices in order to understand how elected representatives, government officials and public servants perceive their responsibility in achieving sound accountability and control in the domain of digitalization and algorithmization.

Reflecting on the categories of data work in practice – Susan Oman

As a Lecturer in Data, AI and Society at the UK's University of Sheffield, I teach our future data scientists (broadly defined) the ethical, practical, environmental, and social limitations and ramifications of data practices. The process of our discussions in the co-authoring of this chapter has led me to reflect on how teaching is not only a 'practitioner intervention in data power', but should be considered a data practice, in and of itself. It is a pedagogical intervention in future data work (or, data work futures), through the shaping of reflective data practitioners. This work is fuelled by the aspiration that future data practitioners will not only be able to reflect on the implications of their own work on people and society, but how it connects to the wider, complex ecosystem of data practices. Teaching responsible data sciences is, therefore, a category of data labour and data work.

Teaching through contextualizing data practices by 'following them around' (Oman, 2021c) the wider data ecosystem, makes them less abstract to students, and crucially, almost impossible to conceive of data without them being connected to people and practices. Who does what with whose data, and who is affected, become urgent questions for those who had perhaps not considered these questions before. The aspiration is that cohorts of future data practitioners will broaden in perspective, life experience, knowledge, and understanding – while developing the core skills expected of a data science programme. Thus, potentially changing what a future data practitioner is as the affective capacities of data and their dynamics become clearer through their learning journeys.

My research looks at what I call 'data contexts' (Oman, 2021a): how data work in different contexts and how different kinds of data work for and against people, differently. I have often focused on administrative data that are routinely collected for social good: to understand well-being or inequality, for example. Much of my research aims to understand different aspects of the public sector and its uses of data, and their implications (whether known or overlooked), and I am currently on an AHRC-funded policy placement in the UK government's department responsible for data and AI.

There are, therefore, many 'layers' of data work (Oman, 2021b) that extend far beyond the traditional idea of a data scientist. In my research, I talk to all kinds of people who work with data across these 'layers' of expertise and practice to understand the performativities of data, and how they 're-perform' (Oman, 2021b). One layer might be construed as the expert–policy nexus, ranging from senior people in national statistics offices, civil servants, and experts who guide or advise on data policy. Another layer is the expert in domain contexts, such as those working in publicly funded organizations who collect and analyse data (including those who consider themselves non-expert). I also talk to people who do not work with data, or at least would not identify as doing so, and so consider this layer as the 'everyday understanding' of data. But, of course, we all work with data. Whether in our jobs, or completing a form to receive welfare or registering with a doctor. We are all doing the emotional labour of the dynamic data infrastructure that permeates all aspects of our lives. Crucially, we also all feel the relations between knowledge and power, even if we may not foreground this in this way when talking about what matters most to us.

We decided that this chapter's focus would be on the tensions of identity, context, and knowledge in relation to data power. As such, it speaks to what I have found across my research projects and professional roles that aim to intervene in bad data practices. This leads me to reflect on what a good data practice even is. Those in national statistics offices believe the administrative data that they work with are ostensibly used to inform knowledge on how to improve society. Those in public data roles often do not have the chance to reflect on the ways evidence and knowledge generated from their data practices are used. There are no reflections on the limits to an improved society. Yet these data about population well-being, or inequality, are pivotal in decisions that affect our life chances, livelihoods, and quality of life.

'Following the data' was a familiar phrase in COVID-19 policy communications for the UK government. If you follow data back centuries, you find it has long been used to track the health and wealth of society. Even when the assumptions that have underpinned over 200 years of social science, statistical and policy work are called into question at the practitioner level, are they questioned enough? Finding ways to follow data practices in their contexts is one way to reveal how data uses actually operate and what they enact. This critical data practice calls for, and, we hope, enables data practices more widely to become more reflective. It is, therefore, a category of data work that incorporates overlooked data labour practices in order to intervene in data power.

Data work by activists, journalists, and community – Joanna Redden

Throughout this chapter we have been drawing on our research and teaching to argue that data practitioners work in a range of roles that extend beyond data science. Our goal is to expand how we think about data work and who is doing that work. The previous sections of this chapter have focused on the data work being done by civil servants, statisticians, economists, international development workers, teachers, and elected representatives. As noted by Oman, there are 'many layers of data work' across a range of contexts. As suggested by Tichenor, in a world so mediated by data practices we are all becoming data practitioners as we confront the ubiquitous collection of our data as datafied ways of knowing influence the ways we are scored, provided or denied opportunities, targeted, and are able to access needed services.

As detailed by Hadley, data practitioners are influenced and often limited by the contextual forces they work within. Renkema and Muis raise important questions about power imbalances, our abilities to know and influence data practices, as well as the extent to which our democratic institutions are at present equipped to ensure accountability, effective oversight, and prevent harm while protecting human rights.

The work I have been doing with my Data Justice Lab colleagues adds to who we might think of as data practitioners by demonstrating the key data work being done by activists, civil society organizations, members of affected communities, and journalists. After previous work mapping and analysing the uses of ADS across public services as well as discussing data harms and resistance, our recent research involved investigating where and why government agencies had decided to cancel their use of ADS in Australia, Canada, Europe, New Zealand, the United Kingdom, and the

United States (Redden et al, 2022). We identified 61 systems cancelled and that a range of factors had led to cancellation. One of our central findings is that nearly half of the systems cancelled had been subject to civil society critique and mobilization as well as critical media coverage. In many cases activist mobilizations and media reporting made the existence of systems in use visible as well as the impacts such systems were having on people. We also found that nearly one third of cancelled systems were stopped as a result of legal action.

This work demonstrates that contrary to the view of some that ADS are neutral and apolitical, their use raises significant concerns related to rights and impacts. Through case study investigations involving interviews and document analysis we stress that there are a range of factors leading to cancellation, often working together. In addition to the factors listed earlier we also found the concerns of civil servants about systems not being as effective as promised as significant, as well as concerns about potential discrimination and bias. In combination, our study of cancelled systems reinforces previous work that stresses that technologies are sites of struggle and power, and that there are competing values, politics, and visions informing decisions about how technologies are employed (Eubanks; 2018; Benjamin, 2019).

Some examples of the key role played by activists, affected communities, civil society organizations, and journalists include work done in Los Angeles in response to police uses of predictive policing. The Stop LAPD Spying Coalition have been credited with researching the impact of police uses of these ADS systems. Their research demonstrated how systems called PredPol and Laser disproportionately negatively affected Black, Latinx, and other people of colour. Their research was featured in media coverage. Stop LAPD Spying Coalition community mobilizing work included calling on the Office of the Inspector General to review the systems. The Inspector General's report pointed to a range of problems. Laser was suspended in 2019 and PredPol in 2020. Similar community mobilizations, critical reporting, and litigation in Australia in response to its automated fraud detection system dubbed Robodebt, led to that system being cancelled in 2019; a judge ruled the system unlawful and a class action lawsuit was settled for $1.2 billion. These are only two examples of community mobilization discussed in our report. They demonstrate the kind of work being done to challenge the distancing and reductive effect of ADS. They also demonstrate the importance of care and community and present lessons in how data literacy and mobilization can be done together.

Conclusion

In combination, our research points to the importance of critical interventions into how we think of data practices and who is a data practitioner. Given the power and ubiquity of new and emerging data practices, we stress the importance of recognizing the data work being done across our societies, and in particular the tensions and struggles connected to datafication. While we focus on different areas of data practice, we each see ongoing concerns around governance and what good practice is and should be. Data work, as detailed throughout this chapter, involves contending with the ways contemporary uses of data are built upon longer histories of unjust, capitalist and colonialist ways of knowing and controlling. Tensions emerging around issues of care and connection versus harm and distancing are linked to context-specific limitations but also competing political visions. A politics of care approach stresses the importance of our connections and interdependence with each other and the natural world (Chassmen and Cohen, 2020; Woodly et al, 2021). In terms of data work, this would involve ensuring decision-making about data practices that is historically and contextually informed while centring social, economic, and human rights. In the context of automation, critical refusal has been argued as necessary for decision-making (Cifor et al, 2019; Gangadharan, 2021; Hoffmann, 2021). As our quotidian social, political, and economic practices continue to coalesce with our data practices – a merging that seems particularly rapid with the current exponential development of large language models – it is important for critical data scholars to take stock of the new forms of responsibility and care that arise, while helping construct our potential data futures. This is the critical data work we aim to present as our own practitioner interventions in data power.

DISCUSSANT RESPONSES

Rethinking data practice – Teresa Cerratto Pargman[12]

'Practitioner Interventions in Data Power' written by Hadley Beresford, Iris Muis, Susan Oman, Joanna Redden, Elise Renkema, and Marlee Tichenor is a compelling, timely, and critical piece of collective work on data power (Hepp et al, 2022). It contributes accounts of what data work entails in practice viewed from different practitioners' roles, mandates, and perspectives. Following the data work conducted by civil servants, local elected representatives, statisticians, economists, international development workers, teachers, and activists, readers will find in this chapter a call to question underlying assumptions regarding who a data practitioner is, how a data practitioner becomes one, and what data work entails in practice.

The chapter invites us to rethink what we understand by data literacy and responsibility as data are not generated, shared, analysed, and used only by data scientists (professionals) but also by the citizens in the different roles they occupy in an increasingly datafied society. This particular understanding that 'we all work with data', there are 'many layers' of data work, and 'we are all becoming data practitioners as we confront the ubiquitous collection of our data' is at the core of this collective work that seeks to generate debate about the following questions: What is a good data practice and what should a good practice be? When is someone data literate enough to make decisions about data and algorithms? How can the domain of digitalization and algorithmization be embedded in existing democratic institutional processes? How do data practitioners perceive their capacity to work with data? How do they see or define their responsibilities?

These questions echo some of the enquiries driving our own research work in the project 'Ethical and Legal Challenges in Relationship with Emerging AI-driven Practices in Higher Education', funded by the Wallenberg Foundations in Sweden. In this project, we engage with issues regarding responsibility for and accountability vis-à-vis the student data that are collected, shared, analysed, and stored in the institutional management systems, online invigilator systems, or automated grading systems, fundamentally configuring the communication between students and teachers/institutions. As the authors in the chapter, we also view the need to conduct empirical studies to contribute to critical studies on data that inform how data practices unfold in situ and reflect on the importance of contextualizing data practices, making them less abstract, as well as connecting them to real consequences for different groups of people in society. Such empirical studies on situated data practices are most needed today to strengthen critical arguments contending the 'structural power differentials in society and the work necessary toward dismantling them' (D'Ignazio and Klein, 2019). Drawing on data feminism (D'Ignazio and Klein, 2019), we also call for a critical understanding regarding where AI systems, hungry for student data, should (and should not) be deployed, for whom, and why (Cerratto Pargman et al, 2023).

The chapter also reminds us that far from being 'objective', 'neat', and 'flawless', data are a site of struggle and power, reflecting competing values, especially when data are in the hands of governments and private global corporations. In this sense, the chapter emphasizes the relation between data as power and algorithmic bias by making clear that algorithmic bias mitigation in the public sector is not only a problem of the current limitations of the different types of technical checks available but also of the broader sociocultural, politico-economic context (people's practices) in which these algorithms operate. As such, the chapter convincingly argues for acknowledging data practice as involving not only data scientists but also a large variety of social and political actors with often disparate skills, knowledge, values, and interests. This renewed and fresh understanding of data practices also points to the heterogeneous composition of data practitioners and, thus, to the tensions and conflicting interests emerging among those dealing with, speaking about, and making daily decisions regarding people's data.

In summary, this is an original chapter of interest to anyone studying and working with data, algorithms, and people. It shares food for thought based on concrete, real experiences gained from the field that compels us to revise taken-for-granted ideas about data practices, data practitioners, and the inherent 'tensions of identity, context, and knowledge in relation to data power'.

Questioning data practitioner tropes and the need for diverse practitioner roles and responsibilities in data work – Caitlin Bentley[13]

The chapter, which examines the landscape of data work and its ethical dimensions, is an enlightening contribution to critical data studies. The authors shed light on the complex layers of what constitutes data work and how this is perceived differently by various actors – be it policy makers, data scientists, or those who see themselves as laypersons in the context of data work. The authors make a compelling case for the importance of critical data studies as a form of data work itself, capturing hidden, harmful, or negligent practices with the aim of making the future of data more equitable and ethical. It pushes the reader to not only question what data work is but also what it ought to be, compelling us to consider ethical considerations and power struggles that are usually glossed over.

The chapter opens with a timely question about the urgency of ethical considerations in data practices, which is especially relevant given current debates surrounding AI, automation, and labour transformation. But who is responsible for making data work more responsible and ethical in reality? The authors adopt a critical data studies lens to examine the wide array of practitioners involved in data work. What stands out is the chapter's incorporation of diverse perspectives – often highlighting how those that do data work, may not even realize it or identify as data workers. Indeed, the chapter does not shy away from highlighting the varying degrees of accountability – or lack thereof – among different actors in multiple data ecosystems. These accountability tensions come to bear, the authors argue, when we visibilize data work, who it is done by, and how.

In my own research, as an academic at a UK university, we have been investigating the requisite skill set that is needed to make data work more responsible and reflective. When examining frameworks to build on, one approach that has been quite common is categorizing skills based on roles or professions. This delineation, albeit clear, presented itself with a plethora of limitations, and we opted for a more flexible general skills framework encompassing a variety of technical, professional, and strategic skills that could be fit for purpose. Our resolution to adopt a flexible approach stemmed from our position that data and AI ecosystems are dynamic and rapidly changing. The roles and professions that are relevant today may undergo transformations, branching out into specialties or amalgamating into broader roles. In this flux, we focused on how a rigid role-based structure might foster obsolescence rather than agility. However, the chapter also shows that data ecosystems are vibrant tapestries, often blurring traditional professional boundaries, becoming sites where roles that might not conventionally be identified as data-centric play a pivotal role in shaping data narratives. Beresford rightly points out that algorithmic technologies are created by multidisciplinary teams.

The chapter thus aptly emphasizes that data work is not confined to those who are fully immersed in data roles; it stretches to encompass individuals from different realms who are gradually becoming a part of the discourse. This necessitates capacity to develop self-awareness around data work, listening to others, melding essential data work skills from various professions, thus nurturing a fertile ground for responsible innovation and ethical data practice.

As my interest also relates to this arena in which multidisciplinary conversations happen, I was really hoping that the chapter could have delved deeper into specific methods, processes, or questions that data workers use in their reflections and debates. Tichenor's presentation of three types of reflexivity that sustainable development data practitioners use (epistemic, care-ful, and instrumental), and Oman's 'data contexts' and notion around 'following the data' point to innovative reflection techniques that could be implemented throughout data work processes, whereas Renkema and Muis and Redden's contributions suggest punctual activities that could precede or follow data work. Renkema and Muis, in their work with municipal actors in the Netherlands, question how much data literacy these actors need to be able to reflect and make appropriate decisions, suggesting there may be preceding factors before a reflection process. In contrast, Redden examined the use of automated decision-making in public-sector services for data harms and resistance, presumably using qualitative research methods post hoc. I hope another book can be written detailing all of these methods, describing and laying the roadmap for how critical reflection can be done by whom, when, and how.

Overall, this chapter is a must-read for anyone engaged in data work, be it academically or professionally. It offers rich insights into the multiple facets of data work and poses critical questions that challenge existing norms and practices, all while advocating for more ethical and accountable data ecosystems.

Notes

[1] University of Sheffield
[2] Utrecht University
[3] University of Sheffield

4   University of Western Ontario
5   Utrecht University
6   University of Edinburgh
7   Discussant
8   Discussant
9   Facilitator
10 A community connected by the annual United Nations Statistical Commission (UNSC) and its various working groups, and currently shaped by the context of the Sustainable Development Goals (SDGs).
11 This is from an interview I conducted with a statistician tied to the United Nations, and this is how this person was coded in our project documentation and other publications on the topic.
12 Stockholm University
13 Kings College London

References

Algorithm Watch (2019) *Automating Society: Taking Stock of Automated Decision-Making in the EU.* Algorithm Watch. Available at: https://algorit hmwatch.org/en/automating-society-introduction/

Amin, S. (2006) The Millennium Development Goals: A critique from the South. *Monthly Review*, 57(10): 1–15. Available at: https://monthly review.org/2006/03/01/the-millennium-development-goals-a-critique-from-the-south

Angwin, J., Larson, J., Mattu, S., and Kirchner, L. (2016) *Machine Bias.* ProPublica. Available at: www.propublica.org/article/machine-bias-risk-assessments-in-criminal-sentencing

Balayn, A. and Gürses, S. (2021) *Beyond Debiasing: Regulating AI and its Inequalities.* European Digital Rights (EDRi). Available at: https://edri. org/wp-content/uploads/2021/09/EDRi_Beyond-Debiasing-Report_Online.pdf

Bandola-Gill, J., Grek, S., and Tichenor, M. (2022) *Governing the Sustainable Development Goals: Quantification in Global Public Policy.* Sustainable Development Goals Series. Cham: Springer International Publishing. Available at: https://doi.org/10.1007/978-3-031-03938-6

Bandola-Gill, J., Grek, S., and Tichenor, M. (2023) The rise of the reflexive expert? Epistemic, care-ful and instrumental reflexivity in global public policy. *Global Social Policy*, 0(0): 1–18. https://doi.org/10.1177/146801 81221145382

Bellamy, R. K. E., Dey, K., Hind, M., Hoffman, S. C., Houde, S., Kannan, K., et al (2018) AI Fairness 360: An extensible toolkit for detecting, understanding, and mitigating unwanted algorithmic bias. arXiv [cs.AI]. Available at: http://arxiv.org/abs/1810.01943

Benjamin, R. (2019) *Race after Technology: Abolitionist Tools for the New Jim Code.* Cambridge, MA: Polity Books.

Big Brother Watch (2021) *Poverty Panopticon: The hidden algorithms shaping Britain's welfare state.* Big Brother Watch. Available at: https://bigbroth erwatch.org.uk/wp-content/uploads/2021/07/Poverty-Panopticon.pdf

Callon, M. (2006) What Does It Mean to Say Economics Is Performative? *CSI Working Papers Series*, 5, pp 1–58. École des Mines de Paris. Available at: http://halshs.archives-ouvertes.fr/docs/00/09/15/96/PDF/ WP_CSI_005.pdf

Cerratto Pargman, T. C., McGrath, C., Viberg, O., and Knight, S. (2023) New vistas on responsible learning analytics: A data feminist perspective. *Journal of Learning Analytics*, 10(1): 133–148.

Chassmen, D. and Cohen, J. (2020) *The Politics of Care: From Covid 19 to Black Lives Matter.* New York: Verso.

Cifor, M., Garcia, P., Cowan, T. L., Rault, J., Sutherland, T., Chan, A., et al (2019) Feminist Data Manifest-No. Available at: www.manifestno.com/

D'Ignazio, C. and Klein, L. F. (2019) *Data Feminism.* Cambridge, MA: MIT Press.

Das, D., Faasse, P., Karstens, B., and Diederen, P. (2020) Raad weten met digitalisering – Hoe de gemeenteraad kan sturen op de maatschappelijke impact van digitale technologie (Rathenau Instituut).

Dencik, L., Hintz, A., Redden, J., and Warne, H., et al (2018) *Data Scores as Governance: Investigating uses of citizen scoring in public services.* Research Report. Cardiff University.

Eubanks, V. (2018) *Automating Inequality: How High-Tech Tools Profile, Police, and Punish the Poor.* New York: St Martin's Press.

Fukuda-Parr, S. and McNeill, D. (2019) Knowledge and politics in setting and measuring the SDGs: Introduction to Special Issue. *Global Policy*, 10: 5–15. https://doi.org/10.1111/1758-5899.12604

Galhotra, S., Brun, Y., and Meliou, A. (2017) 'Fairness testing: Testing software for discrimination'. *Proceedings of the 2017 11th Joint Meeting on Foundations of Software Engineering.* ACM (ESEC/FSE'17). https://doi.org/ 10.48550/arXiv.1709.03221

Gangadharan, S. P. (2021) 'Digital Exclusion: A Politics of Refusal'. In: L. Bernholz, H. Landemore, and R. Reich (eds) *Digital Technology and Democratic Theory* (pp 113–140). Chicago: University of Chicago Press. https://doi.org/10.7208/9780226748603-005

Green, B. (2021) Data science as political action: Grounding data science in a politics of justice. *Journal of Social Computing*, 2(3): 249–265. https:// doi.org/10.23919/jsc.2021.0029

Hepp, A., Jarke, J., and Kramp, L. (2022) 'New Perspectives in Critical Data Studies: The Ambivalences of Data Power—An Introduction'. In: A. Hepp, J. Jarke, and L. Kramp (eds) *New Perspectives in Critical Data Studies: The Ambivalences of Data Power* (pp 1–23). Cham: Springer International Publishing.

Hoffmann, A. L. (2019) Where fairness fails: Data, algorithms, and the limits of antidiscrimination discourse. *Information Communication and Society*, 22(7): 900–915. https://doi.org/10.1080/1369118X.2019.1573912

Hoffmann, A. L. (2021) Even when you are a solution you are a problem: An uncomfortable reflection on feminist data ethics. *Global Perspectives*, 2(1): 21335. https://doi.org/10.1525/gp.2021.21335

Independent Expert Advisory Group on a Data Revolution for Sustainable Development (IEAG) (2014) *A World that Counts: Mobilising the Data Revolution for Sustainable Development*. United Nations. Available at: www. undatarevolution.org/wp-content/uploads/2014/11/A-World-That-Cou nts.pdf

O'Neil, C. (2017) *Weapons of Math Destruction*. London: Penguin.

Oman, S. (2021a) *Understanding Well-being Data: Improving Social and Cultural Policy, Practice and Research*. London: Palgrave Macmillan. Available Open Access at: https://link.springer.com/book/10.1007/978-3-030-72937-0

Oman, S. (2021b) Re-performance: A critical and reparative methodology for everyday expertise and data practice in policy knowledge. *International Review of Public Policy*, 3(3). Available Open Access at: https://journals. openedition.org/irpp/1833

Oman, S. (2021c) *Following the Data*. [podcast]. 17 September 2021. Available at: https://well-beingdata.com/films-on-player-sheffield-ac-uk/

Orr, W. and Davis, J. L. (2020) Attributions of ethical responsibility by Artificial Intelligence practitioners. *Information Communication and Society*, 23(5): 719–735. https://doi.org/10.1080/1369118X.2020.1713842

Oswald, M., Grace, J., Urwin, S., and Barnes, G. C. (2018) 'Algorithmic risk assessment policing models: Lessons from the Durham HART model and 'Experimental' proportionality. *Information & Communications Technology Law*, 27(2): 223–250. https://doi.org/10.1080/13600834.2018.1458455

Redden, J., Brand, J., Sander, I., and Warne, H. (2022) *Automating Public Services: Learning from Cancelled Systems*. Carnegie UK Trust and Data Justice Lab, October. Available at: www.carnegieuktrust.org.uk/publications/aut omating-public-services-learning-from-cancelled-systems/

Saith, A. (2006) From universal values to Millennium Development Goals: Lost in translation. *Development and Change*, 37(6): 1167–1199. https://doi.org/10.1111/j.1467-7660.2006.00518.x

Woodly, D., Brown, R. H., Marin, M., Threadcraft, S., Harris, C. P., Syedullah, J. and Ticktin, M. (2021) The Politics of Care. *Contemporary Political Theory*, 20: 890–925. https://doi.org/10.1057/s41296-021-00515-8

6

Critiques of Data Colonialism

Benedetta Brevini, Irene Fubara-Manuel, Clément Le Ludec,
Jakob Linaa Jensen, Andrea Jimenez,[1] and Jo Bates[2]

Introduction

In this chapter, we critically interrogate the concept of data colonialism developed by Couldry and Mejias (2019). For them, data production activities occur casually as Internet users engage in social activities on the services of big tech companies. As with other research on digital capitalism, their focus is on the process of appropriation and value creation of data produced by users of social networks (Scholz, 2017). They see it as 'an emerging order' of appropriating and extracting 'social resources for profit through data', without any recognition and valorization of the work behind those data.

Many scholars in the last two decades have investigated the unprecedented consolidation of communication systems, exploring the increasing globalization of capitalist relations and rationales; the concerted pursuit of neoliberal economic policies, and the roll-out of digital technologies with its progressive datafication of everyday life. It is hard to think of an aspect of life that has not been affected by the use of algorithms, automation, and big data, including medicine, education, welfare, voting, dating, communication, law enforcement, warfare, and cyber-security. The focus on data accumulation has occupied centre stage in several recent works, including Fuchs' big data capitalism (2019), Zuboff's surveillance capitalism (2018), and Srnicek's platform capitalism (2017). The common themes of these works are that data has a crucial value, and data collection is highly unequal, tending to replicate existing power asymmetries.

Nick Couldry and Ulises Mejias go further and, drawing on their recent book, *The Costs of Connection* (2019), argue that the specificity of contemporary forms of data extraction is best understood by seeing it as a genuinely new stage of colonialism which is based around the appropriation not (as in historic colonialism) of land, its resources, and the bodies to work those resources, but the flow of human life itself, made valuable in the form of data. This thesis of 'data colonialism' is offered as a characterization not just of what is happening with data in historic colonies, but right across the world, including the Global North. These processes of data appropriation and exploitation are the source of the valuation of major big tech companies such as Facebook or Google through their stock market value. In view of their systemic dimension in terms of the number of users and as 'lead firms' structuring the web, these companies have been described as 'sovereign', encroaching on competences traditionally reserved to the state. In this chapter, we each present a critique of Couldry and Mejias's arguments about the emergence of a new stage of colonialism.

Fubara-Manuel argues that before data capture became computational, it was colonial. They interrogate early scholarship that introduces the metaphor of computation as colonization linking this history to contemporary concepts such as 'data colonialism'. Fubara-Manuel argues for scholarship and research that examines colonization as a tangible and pre-existing system of oppression with a violent data legacy that is continued through computation. They posit that computational methods of data capture, especially in post-colonial relations such as global migration, are a second-degree extraction of formerly colonized people that should not be taken for granted in any intersecting analysis of data and colonization.

Le Ludec goes on to suggest that 'data colonialism' is a fundamental misconceptualization. Using the example of a fieldwork conducted in Madagascar, he argues that Internet users who are allegedly subject to data colonialism cannot be treated in the same way as Madagascar's data producers. Rather than talking about data colonialism, he argues, we should thus be talking about 'data extractivism' or 'digital colonialism'. He argues that these latter concepts better account for the inscription of the digital industry into old-fashioned colonial oppression mechanisms.

Brevini then joins the discussion to argue that, rather than a fundamental misconceptualization, there is a gap in the data colonialism concept with regard to the environment. She argues that an intervention is needed in the data colonialism discourse that introduces an environmental justice critique that takes into account the outstanding environmental harms of data-driven communication technologies development. In all their variety of forms, communication technologies rely on large swathes of land and sea, greenhouse gas-emitting machines and infrastructures that deplete scarce resources in their production, consumption, and disposal. They require increasing amounts of energy, water, and finite resources (Brevini, 2020, 2021). She draws the reader's attention to the diverse ways in which data technologies, including energy-intensive artificial intelligence (AI) models and data centres have environmental impacts that go far beyond their carbon footprints and can only be grasped by exploring the global production/ supply chain and life cycle of communication technologies.

The chapter ends with Linaa Jensen returning to Fubara-Manuel's claim that 'before capture became computational, it was colonial'. He claims that colonial logics of exploitation can be found in the earlier historical context of medieval feudalism. Medieval society, created with strong hierarchies, was based not only on exploitation but also on regimes of visibility and public shaming. In line with his recent book, *The Medieval Internet* (2020), he argues that the contemporary public sphere, mediated by digital technologies, resembles the medieval village with mutual surveillance and mechanisms of public shaming such as pillories and other public punishments. Today, he argues, such regimes of shaming are digitally mediated but contribute to uphold the same regimes of dominance on a micro and macro level as found in medieval societies.

Through the insights in these four contributions, we therefore aim to make a contribution to existing debates about the colonial dynamics of the current era, opening up new pathways for understanding and conceptualizing contemporary logics of extraction and capture.

A second-degree extraction: on capture and colonial violence – Irene Fubara-Manuel

Early scholarship on the role of technology in reorganizing labour and occupational communities deployed a metaphor of computation as colonization (Agre, 1995). This metaphor has continued through to contemporary concepts such as 'data colonialism' that connect data capture to capitalist extraction (Thatcher et al, 2016; Couldry and Mejias, 2019). A broad application of this metaphor does not account for the tangible effects of colonization as an existing system of oppression that pre-dates modern computation. In reducing colonization to a metaphor, the violent history of colonization and its continuation in modern computing are not adequately interrogated.

Capture is one of the two models of data privacy according to Agre (1994). Instead of the older model of surveillance that relied on vision – for instance, state-sanctioned CCTV recording the lives of citizens – smart devices track and capture the everyday minutiae of workers and consumers. For Agre (1994: 112), 'capture is never purely technical but always socio-technical in nature' in that it incorporates pre-existing political ideology into categorizing activities and requires imposition of technologies that change systems of meaning for a group of people. It is this computational imposition that Agre (1995) defines as colonization. However, computational imposition and extraction on their own do not equate to colonization. The digitization of the workflow of a border agent, for instance, is not the same as the systemic subjugation of a people and their culture. As a system of power, colonization has its own logic of imposition that is rooted in violence and dehumanization.

Colonization is a form of 'domination and submission' that instrumentalizes and appropriates indigenous people and their land – 'colonization = "thing-ification"', according to Césaire (2000: 42). Before computers could reduce human activities and 'thing-ify' them as profitable data, colonial apparatuses had mapped, dissected, and categorized indigenous people and their land as commodities. Recall the early development of biometric technologies with the handprint of the Bengalese road contractor, Rajyadhar Kōnāi. William Herschel captured Kōnāi's body, as alternative to a signature on a deed. Herschel then sent this handprint to Francis Galton (Pearson, 1914), who would later establish the science of fingerprinting. By placing Kōnāi's handprint in the archives of colonial violence, capture becomes a symbolic dissection of a colonial subject. Kōnāi's handprint becomes a synecdoche – biometric data stands in for the whole person. The capture of this print, not only reduces Kōnāi, but also commodifies his body and labour as property of the British Empire. In Kōnāi's handprint is a symbol of the colonial imperative of transparency and appropriation of indigenous people as raw data (Fubara-Manuel, 2019).

More than a century after Kōnāi's handprint, biometrics are widely applied in migration in the enforcement of borders and the categorization of border crossers. The pattern of thing-ification remains the same. In the aftermath of British colonization, racialized citizens from former colonies face more scrutiny at the UK border (Cellan-Jones, 2020; McDonald, 2020). They are required to produce more data to verify their identities. The heightened scrutiny can be traced back to visa impositions placed during the 1960s to 1980s, as a response to the problem of 'coloured immigration' (Fubara-Manuel, 2020; Patel, 2021). In the 1940s and 1950s, their parents might have been able to live and work in the UK using their British colonial passports that gave them rights to citizenship. The visa impositions of the 1960s and 1980s meant that migrants from former African and South Asian British colonies submitted more data to cross the UK border. As with Kōnāi's handprint, placing modern visas in the archives of colonial violence, there is a continuation of submission. In addition to the submission of bodies, land, and culture, there is a submission of data.

While it is important to place the capture of migrants in the aftermath of colonization and empire, it is also necessary to study the specific changes digital technology contributes to the oppression of formerly colonized people. Big data as a contemporary technological development, means that one body part does not only stand in for a person, but that one migrant stands in as a prediction for a group of 'suspect nationalities' or preferred travellers, as in the UK's streaming tool algorithm (Joint Council for the Welfare of Immigrants, 2020). International biometric databases such as the

EURODAC (European Asylum Dactyloscopy Database) store the captured fingerprints of asylum seekers in the EU to keep their bodies tethered to the border in spite of their physical location (Amoore and Hall, 2010; European Commission, 2022).

Before capture became computational, it was colonial. The key intervention in this chapter is a call to acknowledge and study the form of second-degree extraction that further reduces thing-ified colonial subjects to 'suspect nationalities', high-risk border crossers, or offenders. As with capitalist expansion, colonization reduces people to data and profit. Separate from capitalist data extraction, there is no feigning of consent, there are no terms of service, but only violence and dehumanization. Socio-technical analysis of the pre-existing thing-ification of indigenous people will further uncover the historical methods of commodification that are not only mimicked but deployed and perpetuated by state-owned and commercial technology companies. It would also contribute to an understanding of indigenous modes of resistance against colonial and computational capture.

Behind data production for AI: old-fashioned post-colonial processes – Clément Le Ludec

For Nick Couldry and Ulises Mejias (2019), today's 'datafication' – or turning the world into data – can be compared to the social quantification of the 19th century in that it allows individuals to be categorized (p 119). Similarly, today's cloud firms are extending yesterday's imperial logic in their quest to appropriate data. This extraction of value is also facilitated and extended by the possibilities of coordination made possible by digital technology. Finally, these players control the logistical infrastructure, which gives them some power over other firms: owning a large number of data centres enables them to offer data storage and computing power services to other, smaller firms.

However, Nick Couldry and Ulises Mejias's (2019) analysis is based on the implicit assumption of a digital specificity that would allow this colonialism of data, an assumption that is questionable insofar as (neo)colonial logics also persist in analogue or non-digital environments. Thus, the question remains whether data colonialism simply refers to the expansion and acceleration of data collection, or whether there are characteristics unique to data-driven colonialism.

Based on our own research on data workers working for French AI companies in French-speaking African countries, and on Latin-American data workers, we argue that it is questionable whether it is really possible to equate the extraction of value from data produced by consumers in Northern countries with the extraction of value from international outsourcing chains for the production of AI. Does the term 'colonialism' apply in the same way? Can we put low-paid workers in Madagascar who spend 40 hours a week 'processing' data on the same level as influencing or Instagram users who post pictures? Outsourced work thus represents a cost in human resources. The 'work' of social network users, on the other hand, enables big tech to earn money directly through advertising. Moreover, can we equate the work experience of these two categories of workers? The former, on the other side of the planet, would rather work to survive, while the latter would benefit from the economic security of the countries of the North.

As part of the DIPLab's project on data workers, we interviewed 147 workers, managers, and directors in ten Malagasy companies (Le Ludec and Cornet, 2023). At the same time, we sent a questionnaire to 296 data workers based in Madagascar. They usually earned between 96 and 126 euros a month, with a huge gap between their pay and that of the team leaders, who are also usually in Madagascar and work in the country, but take home eight to ten times as much (Le Ludec et al, 2023).

We think it might be more appropriate to talk about 'data extractivism'. In fact, the notion of extractivism is already well established in the literature in relation to other industries, such as textiles or mining. First, it refers to the appropriation of value by Northern countries and corporations and the alteration of the source of that value (mines, fields). It also refers to the centralization of economic and social power by the same actors in the countries of the North, based on inequalities that are sometimes pre-existing, especially post-colonial relations. These relationships are sometimes violent (in terms of working conditions) and non-reciprocal (little or no investment in the country).

The work of Aneesh Aneesh (2015) on the communicative and affective work of the employees of service centres, who, thanks to the neutralization of their accent and mimesis, make it possible to communicate with American consumers, is very fruitful to illustrate this post-colonial dynamic. They also show the impact on workers of the adaptation of these centres to the constraints of the American market, particularly in terms of working hours. Still in call centres, this time in the Philippines, Jan Padios (2018) speaks of a process of 'colonial recall', showing how post-colonial relations are being redefined as added value for American companies.

As well as cheap labour, the data industry benefits from a well-educated workforce – most have been to university and speak fluent French, learnt at school, online or in Institut Français classes. Founded in 1883, the Institut Français was originally set up to extend imperial power through language to colonized populations. Workers are thus encouraged to develop the colonial skills needed for the export-oriented data and outsourcing industry. This is also where we feel the 'extractive' dimension is more fruitful, as the remnants of the old colonial order are now being joined by new players who are shaping the acquisition of skills and access to the labour market for young Malagasy workers in Antanarivo.

Antonio Casilli (2017) proposes another framework for analysing digital labour in the Global South, based on the call for a 'digital decolonial turn' to make visible the invisible labour performed by marginalized workers. An interesting contribution is the way the author defines digital labour platforms: on-demand platforms such as Upwork or Airbnb, micro-work platforms such as Amazon Mechanical Turk, online social platforms such as Facebook, and 'smart' platforms such as connected object websites. Taking a post-colonial perspective, Casilli (2017: 3943) argues that 'platform capitalism operates by leveraging social constructs of race and gender'. With this in mind, he outlines a number of theories linked to the concept of 'digital colonialism', which focuses primarily on the data imperialism practised by large technology companies.

To extend this latter structural analysis, Casilli argues that the concept of 'coloniality', which refers to 'long-standing patterns of power that emerged as a result of colonialism, but which define culture, labour, intersubjective relations and knowledge production far beyond the strict confines of colonial administrations', is even more accurate than 'digital colonialism' as it also 'recounts the assumptions of dominant discourses while seeking emancipation for all marginalized identities at work' (Casilli, 2017: 3947).

Finally, rather than extending the features of colonialism to the digital oppression of large digital corporations (like in the 'data colonialism' concept), we need to go back to the existing relations of digital colonization. Our fieldwork in Madagascar thus illustrates the direct continuity between colonial institutions and the development of the outsourcing industry. Indeed, the AI sector benefits from a specific policy – the 'tax-free zones' created in 1989 for the textile industry. Since the early 1990s, French companies have been setting up satellites in Madagascar, particularly for the digital publishing industry. These special zones, which exist in many other developing countries, attract investment by offering highly attractive tax exemptions.

Today, of the 48 companies offering digital services in the tax-free zones, only nine are owned by Malagasy citizens, compared with 26 owned by French entrepreneurs. Apart from the situation with formal companies, the sector has developed a practice of cascade subcontracting, with informal companies and entrepreneurs at the bottom of the pecking order, treated poorly and called in when there are labour shortages elsewhere in the sector.

At the end of the AI value chain, workers are struggling with local living standards. Ultimately, they cope with the situation by developing 'colonial skills' which will reversely feed the AI industry with limited long-term benefit for Madagascar. Our fieldwork illustrates why we criticize Nick Couldry and Ulises Mejias's (2019) concept. Hence, we consider that 'digital coloniality' and 'data extractivism' are both concepts that take better account of the institutional and economic mechanisms underpinning the development of the data industry than 'data colonialism' while restoring workers' voices.

An ecological critique of data colonialism: the hidden environmental costs of data-driven technologies – Benedetta Brevini

In this intervention I argue that data colonialism as a framework misses a crucial component of analysis. It needs an ecological critique that places at centre stage the outstanding environmental damage of data-driven communication technologies development. In all its variety of forms, communication technologies rely on large swathes of land and sea, greenhouse gas-emitting machines and infrastructures that deplete scarce resources in their production, consumption, and disposal. They require increasing amounts of energy, water, and finite resources (Brevini, 2020, 2021).

In order to develop an ecological critique of data colonialism, it is crucial to recognize its environmental costs by exploring the global production supply chain and life cycle of communication technologies, from the extractionism characterizing the mining of metals and minerals that are crucial to produce hardware, to the energy costs of hyper-consumption of data colonialism, to the ecological toll of disposal of the devices employed and its consequent electronic waste (e-waste) and e-pollution emergency.

In her research on digital advancements within humanitarian structures, Mirca Madianou (2019: 2) introduces the concept of 'technocolonialism' to examine how the convergence of digital developments, humanitarian structures, and market forces revitalize and reshape colonial legacies. This persistence of colonial genealogies and inequalities is also evident in the global supply chains' data-driven communication technologies, as the extractive nature of technocolonialism extends to the mining of minerals required for digital technologies. For instance, the increasing demand for mineral resources, driven by the adoption of AI, is projected to lead to a 3500 per cent rise in lithium demand in the European Union by 2050 (EC, 2022). Many mining projects in the Global South are located in indigenous territories. The expansion of mining activities can lead to the forced displacement of indigenous communities from their ancestral lands, often without adequate compensation or consent. Additionally, labour conditions in mines can be hazardous and exploitative, with workers facing unsafe working conditions, inadequate pay, and limited access to healthcare and social protections (Smart, 2017).

Moving along the global supply chain, the production of data-driven communication technologies also carries significant environmental costs. A study conducted by the College of Information and Computer Sciences at the University of Massachusetts Amherst (Strubell et al, 2019) quantifies the energy consumption associated with running AI programs. In the examined case of a common AI training model in Linguistics, the study found that it could emit over 284 tonnes of carbon dioxide equivalent, equivalent to the lifetime emissions of five average American cars or approximately 100 return flights from London to New York City. Additionally, the convergence of communication systems that support AI generates its own set of environmental challenges, including energy consumption, emissions, material toxicity, and e-waste (Brevini and Murdock, 2017). To this calculation, it is crucial to add the increasingly worrying water consumption of applications such as Chat GPT, particularly alarming considering the growth and increasing complexity of AI models (George et al, 2023). Data centres, which play a crucial role in the global supply chain, consume an average of 200 terawatt hours (TWh) of energy annually, surpassing the national energy consumption of some countries, such as Iran (International Energy Agency, 2017; Jones, 2018). Furthermore, the information and communications technology (ICT) sector, encompassing mobile phone networks, digital devices, and television, accounts for 2 per cent of global emissions (Jones, 2018). Greenhouse gas emissions from the ICT industry could surpass 14 per cent worldwide by 2040, exceeding half of the current contribution from the entire transportation sector (Belkhir and Elmeligi, 2018). The cooling systems of data centres require substantial amounts of water, raising policy concerns in regions such as the US and Australia that have experienced prolonged droughts (Mosco, 2017). As explained on Google's DeepMind website (Evans and Gao, 2016), one of the primary energy-consuming elements in data centres is cooling. The excess heat generated by servers powering Google services such as Search, Gmail, and YouTube necessitates the use of large industrial equipment like pumps, chillers, and cooling towers. DeepMind proposes Machine Learning as a solution to this issue, which, ironically, is also highly energy-intensive and produces carbon emissions.

At the end of the global supply chain, the disposal of devices becomes a pressing concern. When communication devices are discarded, they contribute to e-waste, presenting challenges for local municipalities in terms of safe disposal. This challenge is so overwhelming that it is often outsourced, and numerous developing economies now find themselves serving as digital dumping sites for more affluent nations, as illustrated by the situation in Kenya (Brevini and Murdock, 2017; Naipanoi, 2021).

In sum, the framework of 'data colonialism' fruitfully highlights a series of alarming social justice concerns, denouncing the power inequalities between the entities collecting and controlling data, the violation of privacy rights and sovereignty rights of those from whom the data is extracted. However, it fails to address environmental justice considerations that are, in fact, a crucial component of social justice as they aim to rectify historical and ongoing inequalities in the distribution of environmental benefits and costs.

The digital medieval village: social media as a new feudalism – Jakob Linaa Jensen

As mentioned by Irene Fubara-Manuel in the first section of this chapter, before capture became computational, it was colonial. As colonialism was a globalized system of inequalities and exploitation, such mechanisms were found much earlier than the colonial era. In the Middle Ages such an economic system was known as feudalism, a term defined by French historian Max Bloch (1961). It denotes a hierarchical system of exploitation where the serfs and peasants were at the bottom of the pyramid. Their work was exploited by the lords in exchange for some kind of protection and stability. On the other hand, the lords themselves were exploited by the kings and nobility through taxes and army conscriptions. In exchange they got land titles and the right to rule those below them. The central good in the medieval economy was labour, just as data is a central good in the platform economy (van Dijck et al, 2018).

At the core of the feudal system were strong logics of visibility and surveillance, organized through spatial practices. Medieval cities were characterized by close-knit communities organized within narrow city walls with ubiquitous surveillance and power hierarchies upheld by public punishments. The penal system was based on public shaming of the offenders, in stocks, wooden horses, or even in the gallows. Punishment was public and spectacular and served to uphold the existing societal order. The socially mediated public sphere similarly combines these two phenomena: visibility and mutual surveillance.

The medieval village forms a powerful metaphor for the logics of sociality and surveillance in the platform economy. Even if not actively commenting or liking, users keep an eye on each other, through what I have elsewhere dubbed as a social sixth sense (Jensen, 2009). Many people write and publish on social media based on anticipated reactions. It is a paradox that the Internet, the ultimate symbol of modernity, transparency, and enlightenment, facilitates logics of enclosure, censorship, and social control.

Social media, like the rest of the Internet, was intended to ease communication and encourage connection, but as it gathered momentum the consciousness of the 'omnopticon' (Jensen, 2007) intruded on the surface of all Internet activities, creating a norm that cripples rather than liberates, making people conform even more to existing norms of what is perceived as mass approved behaviour.

This is particularly striking in the case of what I call 'digital pillories'. In medieval society, the pillory was one of the great instruments of public shaming. Here offenders were exhibited for shorter or longer times for the purpose of public censure. They received public ridicule, outright hatred, and physical acts such as hitting and spitting from the bypassers. The pillory was a strong instrument of social control.

According to French philosopher Michel Foucault (1975), public shaming and punishment disappeared in modern societies and were instead privatized and hidden within the walls of prisons and asylums. However, public shaming as rituals of power, discipline, and the sustenance of inequalities never really disappeared. It still has a presence in modern society, not least in the age of social media and the culture of visibility. As social media are used as a powerful spectacle for celebrities and ordinary citizens alike, they have also proved to be tools for public shaming and what I have elsewhere defined as digital witch hunts (Jensen, 2020).

Witch hunts are often seen as characteristic of medieval society, although they peaked in the sixteenth and seventeenth centuries (Jensen, 2020). Online witch hunts only deviate from their historic predecessors in the method of punishment. After all, the mob does not have formal judicial power. But the rapid and often wrong judgement of 'the social media court of justice' can do real harm to their victims anyway. There are numerous examples of teenagers who have committed suicides because of online bullying, or innocent people who have been driven to suicide by false accusations of crimes. The judgement by the crowd belongs in archaic contexts such as the medieval marketplace. A particularly striking case is the Boston Marathon terrorist attack, where the wrong persons were accused as a result of a social media-led private 'investigation'.

Not all digital pillories are created explicitly with the purpose of public shaming. In recent decades we have seen an explosion of online rating services, evaluating everything from travel agencies and restaurants to teachers and solicitors. The idea is in line with the touted notion that the middlemen – the tourist guide, the critic, the professional – were now unnecessary. Rather than relying on curated advice by experts, everybody can become a reviewer. The idea of trusting fellow consumers and citizens is immediately appealing. Further, in rating portals the expertise is crowdsourced, Wikipedia-style, while the number of reviews increases the reliability, making one odd review less important. The rating regime is closely connected to logics of appropriation and value creation through commodified crowd-sourcing, as discussed in the introduction.

In short, using the feudal logics of the Middle Ages and the Medieval village as an analytical prism for understanding the contemporary platform economy, we can go a step deeper and further back in understanding contemporary logics of platformization, data colonialism, and hierarchical and asymmetric distributions of power.

Conclusion

In conclusion, in our chapter we propose to re-theorize data colonialism by looking at the continuity of practices of appropriating data-related values with older forms of oppression. Whether we are talking about the roots of biometric technologies in Bengalore (India) or the data industry between France and Madagascar, everything suggests that we need to go back to truly colonial or post-colonial inequalities in order to analyse the development of big tech. Jensen's piece even suggests going deeper, using medieval feudal logic to make sense of contemporary logics of platforming, data colonialism, and hierarchical and asymmetric power relations. Our chapter also suggests a number of avenues for further research, for example the issue of the environment is identified by Brevini as a major gap in the work of Couldry and Mejias (2019), who advocates for analysis of environmental costs induced by digital and data-related activities to be integrated into the data colonialism framework.

DISCUSSANT RESPONSE

Unpacking the ambiguity of data colonialism – Andrea Jimenez

In this intervention, I will present my critique of the concept of data colonialism, which is around its vagueness and fuzziness. My contention is that the concept's ambiguity has aided its current popularity while also making it difficult to operationalize in current data practices and a further focus on transformational justice. This ambiguity is not just a conceptual problem, it has material implications.

Extensive literature demonstrates how digital and data-driven technologies perpetuate colonial legacies, often within distinct academic fields, including science and technology studies (Harding, 2011), computer science (Philip et al, 2012), philosophy of technology (Wong, 2012; Mohamed et al, 2020), critical data studies (Couldry and Mejias, 2023), communication studies (Casilli, 2017; Madianou, 2019), human–computer interaction (Reynolds-Cuellar, 2021), and AI (Mhlambi, 2020; Birhane, 2021).

This varied body of knowledge has produced concepts describing similar characteristics. Concepts like technocolonialism (Madianou, 2019) and digital colonialism (Kwet, 2019) also advance our understanding of how data-driven digital technologies are being deployed in ways that represent a distinct new stage of colonialism. As well as data colonialism, these concepts place attention on the larger societal power structures, considering who benefits and who suffers from the current applications of data practices (Jasanoff,

2007). For example, in how technology corporations collect data from marginalized populations to profit from them, providing big technological corporations with 'immense political, economic, and social power' (Kwet, 2019: 4). Or how state-led datafication can negatively impact vulnerable communities in post-colonial contexts (Krishna, 2021). Examples also include how humanitarian and international development organizations use digital technologies to improve the efficacy of their services (Madianou, 2019).

The question that arises to me is how such a diverse set of experiences all can be speaking of the aftermaths of colonialism (Quijano and Ennis, 2000; Maldonado-Torres, 2016). Although I understand that the span of time and geographic scope of colonialism has led to a wide range of diverging and overlapping perspectives (Jimenez et al, 2022b, 2022c), I believe it is important to recognize a distinction between 'decolonization' as the process of returning indigenous land, repatriation of resources, and rebalancing of power (Tuck and Yang, 2012); and 'decoloniality', as the work focused on the aftermaths of colonialism, which have resulted in asymmetrical logics, power, and knowledge structures (Maldonado-Torres, 2016). Although both views are interrelated, I see the importance of this distinction because it helps understand Fubara-Manuel's compelling argument in relation to Le Ludec's one. If we adopt Tuck and Yang's prevalent argument that 'decolonization is not a metaphor' as it refers to the repatriation of indigenous land, then we emphasize how data capture pertains to colonized subjects. This specificity implies that the capture and extraction of daily life might not neatly fit into this category. Correctly so. The subjugation and displacement of indigenous people must be viewed as the central legacy of colonialism today.

Nonetheless, we must recognize when data practices and their global supply chains are part of a larger structure of capitalism/modernity/racism/ heteropatriarchy with colonial roots (Grosfoguel, 2011). Colonialism was so extensive and pervasive that it is as much about dehumanization as it is about the oxygen we breathe. Not only is it ubiquitous, but it is also shaping our future, which is why it is also about its current and future environmental costs, as argued by Brevini. And so, in the absence of clear distinction between colonization or coloniality, data colonialism can be vague and fuzzy, a 'catch-all' ideas concept that raises more questions than answers.

The presence of ambiguity has moved our attention away from using the concept in our work. In our research around co-producing the Water Observatory in Lima, Peru (see Hoefsloot et al, 2022; Jimenez et al, 2022a), we noted how the existing data practices and water infrastructures date back to colonial times, when water data and distribution were only directed at more powerful and affluent people (Bell, 2022). We decided to centre the notion of data justice as a way to work towards fairer data practices. This meant that, beyond showing how government data practices were rendering marginalized people invisible, we actively engaged in creating a platform where those who were not included in the official datasets had the option to be so. While a focus on data colonialism would have centred our discussions on what is colonial about this data practice and/or where coloniality begins and ends, a focus on justice required us to go further and actively attempt to eliminate the injustices we observed.

And so, as data colonialism becomes increasingly popular, a stronger push for clarity and conceptual focus is needed. Otherwise it risks becoming another buzzword in the decolonial bandwagon (Moosavi, 2020).

Notes
[1] Discussant
[2] Facilitator

References

Agre, P. E. (1994) Surveillance and capture: two models of privacy. *Information Society*, 10(2): 101–127. https://doi.org/10.1080/01972243.1994.9960162

Agre, P. E. (1995) From high tech to human tech: empowerment, measurement, and social studies of computing. *Computer Supported Cooperative Work*, 3(2): 167–195. https://doi.org/10.1007/BF00773446

Amoore, L. and Hall, A. (2010) Border theatre: on the arts of security and resistance. *Cultural Geographies*, 17(3): 299–319. https://doi.org/10.1177/1474474010368604

Aneesh, A. (2015) *Neutral Accent: How Language, Labor, and Life Become Global*. Durham, NC: Duke University Press.

Belkhir, L. and Elmeligi, A. (2018) Assessing ICT global emissions footprint: trends to 2040 & recommendations. *Journal of Cleaner Production*, 177: 448–463. doi:10.1016/j.jclepro.2017.12.239

Bell, M. G. (2022) Overlooked legacies: climate vulnerability and risk as incrementally constructed in the municipal drinking water system of Lima, Peru (1578–2017). *Geoforum*, 132: 205–218. https://doi.org/10.1016/j.geoforum.2021.02.016

Birhane, A. (2021) Algorithmic injustice: a relational ethics approach. *Patterns*, 2(2): 100205. https://doi.org/10.1016/j.patter.2021.100205

Bloch, M. (1961) *Feudal Society*. Chicago: University of Chicago Press.

Brevini, B. (2020) Black boxes, not green: mythologizing artificial intelligence and omitting the environment. *Big Data & Society*, 7(2). doi:10.1177/2053951720935141

Brevini, B. (2021) *Is AI Good for the Planet?* Cambridge: Polity.

Brevini, B. and Murdock, G. (2017) *Carbon Capitalism and Communication*. London: Palgrave Macmillan.

Casilli, A. A. (2017) Digital labor studies go global: toward a digital decolonial turn. *International Journal of Communication*, 11: 3934–3954.

Cellan-Jones, R. (2020) 'Home office drops "racist" visa decision algorithm', BBC News, 4 August. Available at: www.bbc.com/news/technology-53650758

Césaire, A. (2000) *Discourse on Colonialism*. New York: New York University Press.

Couldry, N. and Mejias, U. A. (2019) *The Costs of Connection: How Data Is Colonizing Human Life and Appropriating It for Capitalism*. 1st edition. Stanford, CA: Stanford University Press.

Couldry, N. and Mejias, U. A. (2023) The decolonial turn in data and technology research: what is at stake and where is it heading? *Information, Communication & Society*, 26(4): 786–802. https://doi.org/10.1080/13691 18X.2021.1986102

European Commission (EC) (2022) EURODAC (European Asylum Dactyloscopy Database) | Knowledge for policy, European Commission. Available at: https://knowledge4policy.ec.europa.eu/dataset/ds00008_en

Evans, R. and Gao, J. (2016) 'DeepMind AI reduces Google data centre cooling bill by 40%', DeepMind. Available at: https://deepmind.com/blog/article/deepmind-ai-reduces-google-data-centre-cooling-bill-40

Foucault, M. (1975) *Discipline and Punish*. Paris: Gallimard.

Fubara-Manuel, I. (2019) Animating opacity: race, borders, and biometric surveillance. PhD thesis, University of Sussex.

Fubara-Manuel, I. (2020) Biometric capture: disrupting the digital codification of black migrants in the UK. *African Diaspora*, 12(1,2): 117–141. https://doi.org/10.1163/18725465-01201002

Fuchs, C. (2019) 'Karl Marx in the Age of Big Data Capitalism'. In: C. Fuchs and D. Chandler (eds) *Digital Objects, Digital Subjects: Interdisciplinary Perspectives on Capitalism, Labour and Politics in the Age of Big Data* (pp 53–72). London: University of Westminster Press.

George, A. S., George, A. H., and Martin, A. G. (2023) The environmental impact of AI: a case study of water consumption by Chat GPT. *Partners Universal International Innovation Journal*, 1(2): 97–104.

Grosfoguel, R. (2011) Decolonizing post-colonial studies and paradigms of political-economy: transmodernity, decolonial thinking, and global coloniality. *TRANSMODERNITY: Journal of Peripheral Cultural Production of the Luso-Hispanic World*, 1(1). https://doi.org/10.5070/T411000004

Harding, S. (ed) (2011) *The Postcolonial Science and Technology Studies Reader*. Durham, NC: Duke University Press.

Hoefsloot, F. I., Jimenez, A., Martinez, J., Miranda Sara, L., and Pfeffer, K. (2022) Eliciting design principles using a data justice framework for participatory urban water governance observatories. *Information Technology for Development*, 28(3): 617–638. https://doi.org/10.1080/02681 102.2022.2091505

International Energy Agency (2017) 'Digitalisation and Energy'. IEA: Paris. Available at: www.iea.org/reports/digitalisation-and-energy

Jasanoff, S. (2007) Technologies of humility. *Nature*, 450(7166), Article 7166. https://doi.org/10.1038/450033a

Jensen, J. L. (2007) 'The Internet Omnopticon: Surveillance or Counter-insurgency'. In: H. Bang and A. Esmark (eds) *New Publics With/out Democracy* (pp 351–380). Copenhagen: Samfundslitteratur.

Jensen, J. L. (2009) Fra onlinefællesskaber til onlinenetværk: Facebook som augmentering af den sociale virkelighed [From online communities to online networks: Facebook as an augmentation of social reality]. *MedieKultur: Journal of Media and Communication Research*, 25(46).

Jensen, J. L. (2020) *The Medieval Internet*. Bingley: Emerald Publishing.

Jimenez, A., Hoefsloot, F. I., and Miranda Sara, L. (2022a) 'The co-production of the Metropolitan Water Observatory (MWO) platform'. KNOW Working Paper No. 8. Available at: www.urban-know.com/no-8-know-working-paper

Jimenez, A., Vannini, S., and Cox, A. (2022b) A holistic decolonial lens for library and information studies. *Journal of Documentation*, 79(1): 224–244. https://doi.org/10.1108/JD-10-2021-0205

Jimenez, A., Delgado, D., Merino, R., and Argumedo, A. (2022c) A decolonial approach to innovation? Building paths towards buen vivir. *The Journal of Development Studies*, 58(9): 1633–1650. https://doi.org/10.1080/00220388.2022.2043281

Joint Council for the Welfare of Immigrants (2020) 'We won! Home office to stop using racist visa algorithm', Joint Council for the Welfare of Immigrants. Available at: www.jcwi.org.uk/news/we-won-home-office-to-stop-using-racist-visa-algorithm

Jones, N. (2018) 'How to stop data centres from gobbling up the world's electricity', *Nature*, 12 September. Available at: www.nature.com/articles/d41586-018-06610-y

Krishna, S. (2021) Digital identity, datafication and social justice: understanding Aadhaar use among informal workers in south India. *Information Technology for Development*, 27(1): 67–90. https://doi.org/10.1080/02681102.2020.1818544

Kwet, M. (2019) Digital colonialism: US empire and the new imperialism in the Global South. *Race & Class*, 60(4): 3–26. https://doi.org/10.1177/0306396818823172

Le Ludec, C. and Cornet, M. (2023) 'How low-paid workers in Madagascar power French tech's AI ambitions'. *The Conversation*, 30 March. Available at: http://theconversation.com/how-low-paid-workers-in-madagascar-power-french-techs-ai-ambitions-202421

Le Ludec, C., Cornet, M., and Casilli, A. A. (2023) The problem with annotation: human labour and outsourcing between France and Madagascar. *Big Data & Society*, 10(2): 20539517231188724. https://doi.org/10.1177/20539517231188723

Madianou, M. (2019) Technocolonialism: digital innovation and data practices in the humanitarian response to refugee crises. *Social Media + Society*, 5(3): 2056305119863146. https://doi.org/10.1177/2056305119863146

Maldonado-Torres, N. (2016) 'Outline of ten theses on coloniality and decoloniality'. Frantz Fanon Foundation. Available at: https://fondation-frantzfanon.com/outline-of-ten-theses-on-coloniality-and-decoloniality/

McDonald, H. (2020) 'Home office to scrap "racist algorithm" for UK visa applicants'. *The Guardian*, 4 August. Available at: www.theguardian.com/uk-news/2020/aug/04/home-office-to-scrap-racist-algorithm-for-uk-visa-applicants

Mhlambi, S. (2020) 'From rationality to relationality: Ubuntu as an ethical and human rights framework for artificial intelligence governance'. Carr Center Discussion Paper Series, 2020–009.

Mohamed, S., Png, M.-T., and Isaac, W. (2020) Decolonial AI: decolonial theory as sociotechnical foresight in artificial intelligence. *Philosophy & Technology*, 33(4): 659–684. https://doi.org/10.1007/s13347-020-00405-8

Moosavi, L. (2020) The decolonial bandwagon and the dangers of intellectual decolonisation. *International Review of Sociology*, 30(2): 332–354. https://doi.org/10.1080/03906701.2020.1776919

Mosco, V. (2017) 'The Next Internet'. In B. Brevini and G. Murdock (eds) *Carbon Capitalism and Communication* (pp 95–107). London: Palgrave Macmillan.

Naipanoi, L. (2021) 'Dumped e-waste threatens Kenyan lives, contributes to global warming'. The Elephant, 6 November. Available at: www.theelephant.info/features/2021/11/06/dumped-e-waste-threatens-kenyan-lives-contributes-to-global-warming/

Padios, Jan M. (2018) *A Nation on the Line: Call Centers As Postcolonial Predicaments in the Philippines*. Durham and London: Duke University Press.

Patel, I. S. (2021) *We're Here Because You Were There: Immigration and the End of Empire*. Brooklyn, NY: Verso Books.

Pearson, K. (1914) *The Life, Letters and Labours of Francis Galton*. Cambridge: Cambridge University Press.

Philip, K., Irani, L., and Dourish, P. (2012) Postcolonial computing: a tactical survey. *Science, Technology, & Human Values*, 37(1): 3–29. https://doi.org/10.1177/0162243910389594

Quijano, A. and Ennis, M. (2000) Coloniality of power, Eurocentrism, and Latin America. *Nepantla: Views from South*, 1(3): 533–580.

Reynolds-Cuellar, P. (2021) 'Decolonial pathways: our manifesto for a decolonizing agenda in HCI research and design'. MIT Media Lab. Available at: www.media.mit.edu/publications/decolonial-pathways-our-manifesto-for-a-decolonizing-agenda-in-hci-research-and-design/

Scholz, T. (2017) *Uberworked and Underpaid: How Workers Are Disrupting the Digital Economy*. New York: John Wiley & Sons.

Smart, S. (2017) Resistance against mining extractivism in Chile. *Critical Planning*, 23.

Srnicek, N. (2017) *Platform Capitalism*. New York: John Wiley & Sons.

Strubell, E., Ganesh, A., and McCallum, A. (2019) 'Energy and policy considerations for deep learning in NLP'. Cornell University, arXiv:1906.02243.

Thatcher, J., O'Sullivan, D., and Mahmoudi, D. (2016) 'Data colonialism through accumulation by dispossession: new metaphors for daily data'. *Society and Space*, 34(6): 990–1006. Available at: https://doi.org/10.1177/0263775816633195

Tuck, E. and Yang, K. W. (2012) Decolonization is not a metaphor. *Decolonization: Indigeneity, Education & Society*, 1(1), Article 1. https://jps.library.utoronto.ca/index.php/des/article/view/18630

van Dijck, J., Poell, T., and de Waal, M. (2018) *The Platform Society*. New York: Oxford University Press.

Wong, P.-H. (2012) Dao, harmony and personhood: towards a Confucian ethics of technology. *Philosophy & Technology*, 25(1): 67–86. https://doi.org/10.1007/s13347-011-0021-z

Zuboff, S. (2018) *The Age of Surveillance Capitalism: The Fight for a Human Future at the New Frontier of Power*. New York: Profile Books.

7

Environmental Data Power

Patrick Brodie, Monika Fratczak, Olga Gkotsopoulou,
Monika Halkort, Rolien Hoyng, Paul Quinn,
Eric Nost,[1] and Jo Bates[2]

Introduction

Data practices are at the heart of contemporary understandings of environmental change, yet the infrastructures that underpin them are also increasingly understood to have their own significant environmental impacts. In this chapter, the contributors address the issue of the relationship between data and the environment, through explorations of environmental sensing, data–driven representations of climate change, and the environmental impacts of data centres. In the contributions, we are interested in exploring the different entanglements of data and the environment. Specifically, we look at how power dynamics come into play along the journey from environmental sensing practices, through to data modelling and representation, and data storage and processing in data centres.

Across each of the chapter sections, co-authors variously engage with the contours of data, their diverse physical manifestations, their representations, their affects, and the epistemologies they materialize, reproduce, and map onto in environmental and planetary contexts. We ask: how does 'environmental data power' as a concept operate across each of these registers? To answer this question, we articulate not only how data power takes shape in relation to and via environmental contexts, but also how data technologies, economies, and practices organize and make legible the planetary through these particular affects, epistemologies, and geographies.

The first section by Brodie examines how datafication and its relations to the environment extend beyond how the environment is understood through digital technologies – it is also increasingly organized by the material infrastructures required to sustain digital economies. Data centres represent concentrated sites at which the data relations of digital capitalism coalesce and take shape, whether in the form of cloud computing or the structural organization of emerging energy systems. Energetic data power, as Brodie's section articulates, demonstrates how multinational tech companies are grafting digital business models onto energy transitions, exacerbating inequalities within the regimes and infrastructures of decarbonization at the data/energy nexus. Against this, scholars and activists need to understand struggles around data power as at the same time resource struggles at the cutting edge of new material extraction and distribution networks – from data centres to smart grids to wind farms.

We then move on to Halkort's piece which speaks most strongly to the material entanglement of digital infrastructures and their wider social and natural environment that runs through this chapter. What she shows is how the ever more pervasive architecture of environmental sensors and Earth observation technologies at sea create new conflicts over how to distribute solidarities between humans and non-humans in the face of competing pressures – that is, climate change, depleting energy resources, extreme weather events, pandemics, or war. It is in relation to this planetary crisis that the Mediterranean provides an instructive case study to unpack the uneven ways digital infrastructures are mobilized in response to risks of death and extinction and how this uneven distribution is normalized up to a point where it operates silently and imperceptibly in the background. Drawing on the notion of touch as her main analytical device, Halkort wants to bring these hidden, figural tactics back into view and make them addressable as ethical challenges, both for critical scholars and those managing these infrastructures alike.

Next, Gkotsopoulou and Quinn's contribution explores citizen science projects and their relationship to empowerment through information transparency. They consider the ways in which citizen science is shedding a new light onto traditional scientific research practices, comprising the definition of processes, the inclusion of research participants, as well as the designation of interim and end-goals including policy making. In particular, the amount of data generated in citizen science projects with environmental causes is immense, originating from various and diverse sources, and to ensure data quality, a groundwork of bottom-up principles in combination with the provision of transparent definitions and accessible information is necessary. To investigate empowerment through information transparency, they look into the methods deployed by a specific EU project, SOCIO-BEE.

Hoyng then joins the discussion with a section on how carbon metrics quantify emissions and aid algorithmic modelling of the changing climate. Hoyng's essay probes the ambiguous role of uncertainty in metrical, algorithmic representation and problematizes it by unpacking the tension between what she calls the logic of 'if/then' and the mode of 'what if'.

Whereas policy instruments simulate and calculate the consequences of our actions/inaction following the structured logic of 'if/then', the speculative, experimental mode of 'what if' underpins climate modelling as a practice. Data power shows itself in the ways in which metrics construct, produce, enable, and constrain, while they seemingly only measure their empirical referent. Yet, while casting certainty and accuracy, carbon metrics tend to erase uncertainty. While precise calculation enables us to act and address the climate crisis in certain ways, this comes at the expense of accounting for the real challenge, namely acting in the face of uncertainty.

Our final section, by Fratczak, focuses on climate change data visualization (data vis) as a case study that offers a means of exploring how graphs, charts, and maps can encourage people to think – and feel – differently about urgent social issues and mobilize them to act. It centres around the power of data vis in climate change communication and challenges the prevailing assumptions that data vis primarily conveys knowledge and rational arguments. The essay advocates for the recognition of emotions as a crucial element in engaging with data vis and mobilizing individuals in datafied democracies. It underscores the importance of inclusive practices in the design of data vis, ensuring that vulnerable groups have equal access to data and opportunities to shape datafied societies.

In putting together these case studies, which variously speak to the power exerted by and through the intersections of data and the environment, we hope to add to a growing conversation about data and environmental justice. Building on and in conversation with the formative and emerging work of scholars such as Jennifer Gabrys (2014), Mél Hogan (2018), Sebastián Lehuedé (2022), Jenny Goldstein and Eric Nost (2022), Alan Irwin (1995), and Janis L. Dickinson and Rick Bonney (2012), we show in this chapter that environmental data power is also a tool that can be wielded by communities, activists, and political groups towards environmental justice, whether that means disrupting powerful circulations of digital technologies or identifying key points at which to intervene and even seize data power. As environmental justice becomes a moral imperative for scholars of digital media and technology in the midst of deepening climate crisis, so must we align with the principles and actions of data and environmental justice (see also Vera et al, 2019). Utilizing these practical tools and building coalitions across data and environmental justice struggles, we can identify sites through which to build transformative relations of environmental data power.

Energetic data power – Patrick Brodie[3]

Funding disclosure: UCD Ad Astra Starter Fund
In this short excerpt, I will reflect on what I call the 'energetic data power' becoming embedded in infrastructural systems in Ireland at the intersections

of data and renewable energy. Most of my recent research has looked at the data centre industry, and specifically its huge expansion in Ireland (Brodie 2020a, 2020b, 2021; Bresnihan and Brodie 2021, 2023). Sharae Deckard, an environmental humanities scholar, expanding on the work of Jason Moore, argues that the financial mechanisms that have attracted and maintained multinationals here mean that FDI 'organizes nature' in Ireland (Deckard, 2016). She was speaking financially, and via spatial and property development, but in the case of big tech, their infrastructures metabolize with the wider infrastructural systems in which they rely and partake. Data centres act as tools and infrastructures of resource-making in a number of ways – from the types of data they process and the ways they do it in the refining of data-as-resource via data extractivism (see Taffel, 2021), to the physical, mineral, water, and energetic inputs required to construct and run them (see Hogan, 2015; Lally, 2019), which are extracted from land and atmospheres surrounding the facilities and via global supply chains that extend far beyond Ireland.

Their primary resource impact and operational capacity, though, is through electricity use. In 2021, data centres used 14 per cent of the country's grid capacity on a daily basis, as compared to the 12 per cent used by all rural homes combined (see Carswell, 2022). This obviously points to huge inequalities in how energy systems are being used and developed here, especially during an interlocking energy crisis and low-carbon energy transition. The resource-intensivity of data use is largely eaten up via these facilities, whether in the form of what they draw directly from the grid, the battery backups that they charge (also from the grid, but stored in lithium batteries), or the diesel-powered generators they keep on-site in the case of grid failure. All this while folks struggle to afford heat in their homes.

But none of this is to mention the ways in which data centre providers are posing themselves as essential to grid transformation in more entangled ways, demonstrating the backdoor privatization of aspects of utility systems – delivering refined resources for lives and livelihoods through infrastructures – that big tech companies are able to graft onto (see Velkova, 2021). Data centres are at the forefront of the transformation of the Irish energy system, including the state incentivization of corporate power purchase agreements to develop wind and other renewable energy. AWS, for example, has purchased power from four wind farms across Ireland, one of which, the Meenbog Wind Farm in Donegal, was the site of a catastrophic peat landslide in 2020, demonstrating the ecological entanglements with rural places apparently far from the site of the ostensible data infrastructure. Microsoft abruptly announced in 2022 that they had purchased 900 MW of renewable energy in Ireland through these schemes (O'Brien, 2022). These corporate power purchase agreements (CPPAs) act as offsets in the absence of direct energy provision for their data centres, essentially also guaranteeing low prices for electricity into the future.

The overwhelming of Dublin's grid was finally acted upon by Eirgrid in 2021, by recommendation of the Commission for the Regulation of Utilities, with a de facto moratorium on new data centre grid connections in Dublin. The recommendations, however, were not to halt data centre growth entirely, and as of late 2022 there were apparently 21 new data centres in planning outside of Dublin (Moss, 2022) and 11 facilities in Dublin bypassing the grid by connecting to gas networks (Swinhoe, 2023). In light of these ongoing issues, data centres are being proposed by state and corporate supporters as potential grid solutions providers due to their on-site energy infrastructures, including batteries, in order to absorb excess intermittent renewable energy capacity – ensuring 'grid flexibility' by absorbing power and selling it back to the grid (Paananen and Nasr, 2021; Roach, 2022). Microsoft in particular has tested this model in Ireland and plans to export it elsewhere in their data centre fleet (Roach, 2022). When you also take into account the cloud-based technologies for forecasting, monitoring, and measuring energy availability and use along the supply chain of the grid, by companies such as Siemens and GE, the implications of big tech involvement in the operation and delivery of these systems introduces difficult questions about the administration of public utilities by multinationals – whether considering the potential for a tech downturn, or simply in the extractive systems designed to maintain the energy status quo via techno-solutionism.

As the above cases demonstrate, there are huge implications for land and infrastructure in these energy transitions, especially in the backdoor privatization and land-grabbing that are coming in this new resource rush for renewables, not to mention the land and resources required for the basic construction and functioning of data infrastructures. By mapping the myriad implications of big tech's role in Ireland's climate-driven energy policies, we can better understand how territories and environments are being enrolled into these emerging systems, even apparently far from typical imaginations of data-intensivity. Each of these sites can also crucially be understood as key points of resource struggle around the expanding and transforming relations of data power.

Just sense: on the tactility of vision in the Mediterranean Sea – Monika Halkort

The combined impact of climate change, loss of biodiversity, industrial waste, and noise pollution have established the world oceans as a critical platform for anticipating risks of premature deaths and extinction. It transformed coastal waters and the high seas into highly instrumented spaces, where the boundaries between social, technical, and biological intelligence, sensory registers, and communication are becoming ever more blurred (Gabrys, 2019). An extensive array of drifting sensor points, underwater gliders, and Argo floats are currently roaming the seas for the purpose of observing rising sea levels, oceanic temperature, salinity, and ocean currents. They are supported by coastal webcams, Earth observation satellites, and remotely operated underwater vehicles (ROVs, AUVs) or hydrophones, that enable scientists to monitor and document critical changes in the marine ecosystem in next to real time (Gabrys, 2019).

Increasingly these environmental platforms are also mobilized for military purposes and maritime surveillance, further fuelling logistical fantasies of rendering the sea knowable, governable, and transparent. Both NATO and the EU coastguard and border security agency FRONTEX draw on a wide spectrum of real-time tracking and monitoring devices, including the Automated Vessel Identification System, high-frequency radars, underwater sonars, and satellite imagery, to create a matrix of situational awareness for policing the Mediterranean Sea.

In my previous work (Halkort, 2021, 2022), I have shown how this convergence of military and scientific intelligence implicates planetary infrastructures of Earth observation in racialized politics of risk and securitization in the face of competing pressures. In the short space available here I want to emphasize the material agency of instrument platforms in this process, drawing on the notion of touch to unpack how environmental sensors configure oceans, environments, and bodies into 'matters of concern'.

To raise the question of touch in relation to sensory media, is not to insist on the idea of touch as privileged or unmediated access to the world imbued in the tactility of the body or the skin, but rather to make room for interrogating the inherent ambivalence of sensibilities afforded by technical mediation that both open up new possibilities for engaging with neglected perceptions while at the same time allowing for new modes of abandonment and exclusion, silencing, and erasure in our engagement with non-human others and worlds.

A brief look at how machine sensors read and interact with natural environments may help to explicate this point. Two of the most important measures in climate research are temperature and salinity – the degree of salt in water. Both indicate the density of water, which regulates how ocean currents circulate from the tropics to the poles critically affecting global warming and climate change (NASA, 2009). To measure salinity concentrations, the subatomic particles of the seawater (electrons) interact with electrodes on sensors, using the electromagnetic radiation of sea surfaces and electrically charged particles of the water (ions) as their primary medium (Aquaread, nd). The drop in voltage between the two sensor electrodes gives a measure of resistance, indicating the concentration of ions in the water, which can be converted into a measure of salinity. The higher the concentration of ions, the higher the salinity of the water (Aquaread, nd). Hence, what sensors in fact measure is not the amount of salt in the water as such, but the electromagnetic activity running through it, converting it into numerical data that can be processed in machine-readable form.

The ways ocean sensors communicate with their environment powerfully evokes the physicality of touch as a mode of contact and connectivity in cross-modal encounters, where seeing/sensing is never separate from direct engagement with the materiality of objects or environments observed. As Karen Barad (2012: 206) reminds us, from the viewpoint of quantum physics, touch is above all a figure of repulsion and withdrawal – the effect of electromagnetic interaction between particles communicating at a distance, pushing each other away. Hence, what we actually feel, in a touching sensation, is not the caressing sensation of a beam of sunlight or a body, whose touch we may seek, but rather the electromagnetic repulsion between the electrons of the atoms that make up the human body or the entities we touch. Touch in this sense always involves both a connection and a separation – 'a spacing in contact' (Nancy cited in Butchart, 2015: 223) – where the boundaries that hold entities in communicative relation are simultaneously experienced and marked.

From a feminist viewpoint these 'spacings' carry ethical resonance (Haraway, 2007; La Puig de Bellacasa, 2017). They demand a skilful recognition of the multifaceted risks, vulnerabilities, and collective reciprocities at stake in the ability to touch and being touched by human or non-human others and to remain curious about what happens in the contact zones established by touch (Haraway, 2007). Relayed back to the tactile vision afforded by remote sensors this implies attending to the multiple gaps, blind spots, and misrecognitions that come with the expansive field of sensibility afforded by machine vision as they convert lived and embodied phenomena into data signals and machine-readable forms.

In the specific context of the Mediterranean, as I have shown (Halkort, 2021, 2022), these perceptual possibilities are currently mobilized in a highly selective and exclusionary manner. This has created a situation where technologies, originally designed for border protection and maritime surveillance, are put to use for the protection of endangered species, while thousands of migrants are left to die in the same waters, without anyone noticing or responding to their call. This split recognition of risks and emergencies is indicative for the ways digital infrastructures enrol new spaces for validating and evaluating life on the level of data proxies, as a direct result of how they are modelled in data and how they are algorithmically codified.

Fair environmental data power – Olga Gkotsopoulou[4] and Paul Quinn[5]

Funding disclosure: Horizon 2020 project SOCIO-BEE (Wearables and droneS fOr CIty Socio-Environmental Observations and BEhavioral ChangE), Grant agreement ID: 101037648

From the colossal citizen science projects organized around the globe by NASA[6] to smaller, local or district-based initiatives, more and more research depends on the collection and processing of data achieved through citizen science channels. This is particularly true for data relating to the environment within the aim to better inform environmental policies and laws, enhance scientific outcomes for further research, and study and promote scientific literacy and innovation as well as science communication.

In this brief passage, we look into the field of environmental observation and action which includes several diverse citizen science initiatives. We take as an example, a selected EU citizen science project for environmental action, specifically with respect to air quality (SOCIO-BEE).[7] The citizen scientists will be encouraged to collect data about air quality in their urban neighbourhood, to observe the increase of pollutants in the air or the spread of smog, so as to raise awareness within the local communities and inform actions at formal level. Those different types of data coming from citizen scientists, will feed into further research; this research will subsequently feed into specific scientific outcomes; and the outcomes will feed into evidence-based decision–making at municipality level, with sustainable impact.

Citizen engagement is a top priority in the European Union (EU)'s agenda, to enhance community involvement and participation in policy making through the generation, fair sharing, and processing of data. To establish a data-sharing 'culture' and improve data infrastructure, the EU calls, through its 2020 European Strategy for Data, for more sustainable data economy models, featuring data altruism schemes. Based on it, it invests heavily in data law initiatives with the review or repealing of older legal instruments and the negotiation and adoption of new.

Part of citizen engagement is citizen science; in other words, the participation of the public in scientific research, either with the help of, or outside, institutions traditionally regarded as scientific. That said, one first challenge to overcome is that even though there have been many efforts to define citizen science, there is not one commonly agreed definition. In line with the definition included in the *Oxford English Dictionary* in 2014, citizen science is to be understood as 'scientific work undertaken by members of the general public, often in collaboration with or under the direction of professional scientists and scientific institutions'. A more traditional view on citizen science is that of Irwin (1995), which describes citizen science as a movement to democratize science. A more recent approach is that by Groot and Abma, who regard citizen science as 'an umbrella term that applies to a wide range of activities that involve the public in science' (Groot and Abma, 2022). In a nutshell, to highlight the diversity of citizen science initiatives, it is important to mention the different types that exist depending on the role assigned to the citizen scientist (Nierse, 2019), that can be categorized in the following: consultation, collaboration, and control. To add more complexity, there are different types of research projects depending on the level of engagement (Bonney et al, 2015), including data collection projects, data processing projects, curriculum-based projects, and community science ('co-created') projects.

The SOCIO-BEE project includes the study of law, policy, and ethics of data flows in environmental action as an indispensable component by scratching upon the layer of data management, from its conception through research design until its interpretation into a plan and its communication to the citizen scientists in a four-way structure. First a Data Management Plan and an Ethical Commitments Strategy were developed for the project consortium to lay down the threshold legal and ethical requirements and to provide clarity about the data flows and processes from the beginning until the conclusion of the project. Second, a Protocol for Citizen Scientists was developed in cooperation with citizen science experts coming from different fields which sets out the bottom-up principles for the project participants. It functions as a research protocol based upon co-creation and interdisciplinary cooperation, promoting principles of data quality and respectful participation. Third, the knowledge achieved on these matters is being openly shared through the Knowledge Powerhouse for Citizen Science on Law and Ethics,[8] a publicly available repository with a wealth of resources. A final pillar is the provision of accessible information to all research participants, including data subjects, in the project.

All in all, we argue that for power based on environmental data collected or processed within the citizen science context to be fair: (a) light must be shed upon the existing and applicable legal framework on citizen science and research data management as well as the ongoing discussions in the EU; (b) citizen scientists must be emboldened with accessible information with respect to their participation in research processes; and (c) the challenges posed by the ambiguity around the concepts of citizen science, scientific research, as well as new emerging terms such as data altruism, must be acknowledged and taken into consideration when designing and executing citizen science projects, for instance through the standardization of citizen science for high-quality participatory research. An instance of such effort could be the 10 Principles of Citizen Science by the European Citizen Science Association, which provide a groundwork in the form of a *Wegweiser* for citizen science projects in Europe.[9]

Counting carbon, accounting for uncertainty – Rolien Hoyng[10]

Funding disclosure: Direct Grant, The Chinese University of Hong Kong
Carbon metrics quantify emissions in support of algorithmic modelling of the climate, or rather, the interactions between planetary ecosystems and human action. This essay probes the ambiguous role of uncertainty in metrical, algorithmic representation of the climate. I problematize this ambiguity by unpacking the tension between what I call the logic of 'if/ then' and the mode of 'what if'.

From United Nations' COP conferences to carbon tracking apps, responses to the climate crisis rest by and large on quantifying emissions. Carbon acts as a metric (CO_2) rendering different greenhouse gases equivalent in terms of their potential to warm the planet, and so it provides a framework for conceiving pollution (Whitington, 2016). Building on the standardization and quantification enabled by this metric, carbon counting aids the policy instrument of the carbon budget which calculates how much more we can burn, so to speak, before we encounter a threshold – be it 1.5 °C or 2.0 °C. It furthermore underpins simulation in Integrated Assessment Models (IAMs) that incorporate various parameters and proxies representing factors such as the global economy, energy usage, and land and climate systems. These socio-ecological models present algorithmic simulations of the consequences of our actions/inactions, following logics of if/then (Bucher, 2018): 'if this' (policy/behaviour), 'then that' (temperature increase/decrease).

As Brighenti (2018) argues, metrics tend to affect the state of things under observation, and here the metric both enables and delimits prevailing forms of climate action. Yet, despite the appearance of empiricalness and accuracy, carbon counting is often speculative in practice. For instance, not only actual exhaust is counted but also the potential absorption, or intake, by carbon sinks. Such sinks include reforestation projects, involving bacteria experiments as well as technological innovations in carbon capture, even if these technologies do not yet exist or their effectiveness has not been proven so far. In the case of 'green' projects, negative emissions are hypothesized as virtual (not-actualized, not-in-fact-happening) exhaust is rendered as a quantity that can be sold and bought so that, while someone is 'saving' pollution, someone else can buy a licence to pollute. So, carbon trade and offsetting form markets around the right to pollute as a scarce resource.

Moreover, whereas instruments such as carbon budgets and IAMs imply a planet reduced to the metric of carbon and the structured logic of 'if/then', the mode of 'what if', which underpins climate modelling as a practice, probes the planet and human involvement in it in a speculative manner. The notion of a changing climate is constituted by complex models of proxies, speculations, and conjectures. As with Bowker and Star's (1999) infrastructural inversion, a model may seem an object, but look again and all you spot are situated practices of modelling through constant tweaking and recombining different models. The suggestion of an objective system, abiding by logics of 'if/then', dissipates in experimental and speculative practices conducted in the mode of 'what if': what if conditions are as such and such, what if there is a relation, what if weights are as such and such. The problematic of 'what if' becomes, for instance, evident in the use of 'ensemble' techniques in climate change modelling, which rather than deploying only one method of computation combine a range of methods by tweaking either initial conditions, models, or their parameters. The experimental and speculative mode of 'what if' responds to planetary uncertainty and instability – call it 'agency' (Chakrabarty, 2021) – which become increasingly palpable as what are supposed to be unlikely 'black swan' events begin to feel like the norm itself: the unpredictable becomes the expectable. Now, can we talk about a climate budget, which assumes a relatively stable ontology, plus known thresholds and tipping points? Even the notion of 'climate risk', which figures in policy and insurance, evokes a probabilistic ontology in which chance can be calculated, rather than an open (anti-)system consisting in planetary instability, provoked by the climate crisis.

The logic of 'if/then' and the mode of 'what if' are at tension. To understand the politics and consequences of this tension, and the ways in which they play out in particular situations, requires thinking critically about environmental data power. Carbon counting in the context of policy and trade assumes a planet with a rather stable ontology, a system that is more or less fixed, revolving around logics of 'if/then'. Data power shows itself in the ways in which metrics construct, produce, enable, and constrain, while they seemingly only measure their empirical referent. Yet, while casting certainty and accuracy, carbon metrics in the context of the climate crisis often perform a double erasure: they efface the speculative nature of the practices that they support as well as of our understanding of the planet more generally. My point is not to critique data-centric and metric-based epistemologies of the climate crisis for their inaccuracy but to question the idea of accuracy itself, as it erases the uncertainty that primarily qualifies the predicament of climate chaos. The precise calculations in support of trade, offsetting, and risk industries enable us to act and address the climate crisis in certain ways, but at the expense of accounting for the real challenge, namely living with, and acting in the face of, uncertainty.

The power of data vis: emotional responses to climate change data visualizations – Monika Fratczak[11]

There has been a huge rise in the visual communication of data and information in the form of data visualization (data vis) in everyday social life and on social media. Politicians, designers, activists, and campaigners increasingly use data vis in topics ranging from climate change and elections to COVID-19 distribution and social inequalities, hoping that it is an effective tool not only in communicating information to the public and reaching a variety of recipients but also in mobilizing them to undertake the actions they deem necessary. In this context, understanding the role emotions play in engagements with data vis about climate change is especially important because emotions are vital components of making sense of data, as a number of practitioners and scholars argue (such as Kennedy and Hill, 2018; D'Ignazio and Bhargava, 2020; Gray, 2020). Moreover, Nærland (2020) suggests that data vis enables and mobilizes people to function as citizens and take part in political debates or everyday discussions.

However, to date, most existing studies on engagements with data vis have been carried out from a computer science or psychological perspective. These studies have primarily focused on the capacity of data vis to convey knowledge, its efficiency, or comprehension, or its ability to persuade audiences through rational arguments (see Haroz and Whitney, 2012; Haroz et al, 2015; Borkin et al, 2015; Harold et al, 2020). This may be due to an assumption that data, and thus data vis, are objective reflections of knowledge and do not make much of an emotional impact, unlike other visuals, such as emotive images (see, for example, Climate Outreach's project and research, 2021). Therefore, the purpose of data vis is often seen as a tool that allows recipients to independently explore data and draw empirical conclusions. In these processes, emotions do not play a significant role.

In contrast to this assumption, my research considers emotions as a central aspect of social and political experience (Ahmed, 2004; Wahl-Jorgensen, 2018) and a vital aspect of people's engagements with data and data vis (as seen in Kennedy and Hill, 2018). My study offers insights into diverse individuals' emotional responses to data vis about climate change through a sociological lens and considers whether and how these emotional responses may subsequently prompt political participation in different national and geographic contexts. It employs a comparative mixed qualitative methods approach, incorporating visual analysis of 13 data vis about climate change published on Facebook, Twitter, and Instagram, nine semi-structured interviews with ten data vis professionals from six organizations who design, commission, and/or disseminate data vis about climate change on social media, 34 semi-structured interviews and 13 follow-up interviews with diverse social media users who responded to the data vis about climate change disseminated on Facebook, Instagram, and Twitter by Carbon Brief, Climate Science, Greenpeace, and the World Wildlife Fund, organizations from the UK and Poland.

In this study, I argue that data vis can be seen as what I have called an 'emotional repository' of dynamic and complex emotional experiences. These emotions play an important role in mobilizing people to participate in datafied democracies, more often on an individual and daily level, and less frequently on a collective and public scale. However, the ways in which emotions motivate participation can differ depending on national and geographic contexts and other demographic characteristics. Furthermore, unequal emotional engagement with data vis and mobilization or lack thereof, can result from different national contexts and other demographic and social factors. My findings suggest that some demographic, often vulnerable, groups may be discriminated against in data vis practices, as suggested by D'Ignazio and Bhargava (2020). This discrimination may be the reason why some individuals do not emotionally engage with data vis about climate change, do not pay attention to them, and are consequently not mobilized by them.

By providing empirical insights into the emotional significance of data vis, my research challenges two assumptions. First, the assumption that understanding data vis about climate change is the main prerequisite for mobilizing people to act, as many studies focusing on the capacity of data vis to convey knowledge and persuade audiences through rational arguments have shown. Second, the assumption that emotive imagery, such as photos of the negative consequences of climate change, may be more effective means of engagement than data and data vis. While researchers and activists have emphasized the use of emotive images such as photos to engage people with climate change issues (see Leiserowitz, 2006; O'Neill et al, 2013; Metag et al, 2016; O'Neill, 2020), there is no equivalent recognition for data vis. Therefore, higher priority should be given to data vis, with greater recognition of the emotions they can trigger, as these emotions can be the only way for many vulnerable groups to access data and participate (or not) in datafied societies.

Conclusion

In environmental contexts, data have frequently acted as a way of knowing and accounting for environmental changes and transformations, whether towards conservation, development, or other human-centred activities. Across this chapter, however, we have demonstrated that data are far more than a way of understanding and ordering the world – their technologies and infrastructures act upon existing institutions, spaces, and environments. Crucially, this occurs unevenly, and requires thoughtful and critical engagement with sites of data operations in order to re-route data power towards more just and equitable alternatives. Data power is not only wielded and enacted by powerful institutions, it is, and can be further disrupted, challenged, and taken back by a diversity of actors across the contexts studied here.

One provocation we may make here is to ask how we can extract data from their most frequent utility within inequitable and environmentally unsustainable development, profit, and capitalist accumulation applications. As each of these contributions demonstrates or evokes in different ways, data power is frequently instrumentalized towards ostensibly environmental goals that may be at odds with existing and more careful relations to the environment. By extricating environmental data power from the institutions and infrastructures whose strategies have long harmed and degraded the environment and reproduced environmental inequities and injustices across history, researchers and activists are, and have been, able to redirect environmental data power towards community- and justice-centred alternatives to (big) data capitalism and its imbalanced environmental relationships. In this way, we hope that this chapter presents not only critique of environmental data power exerted and exploited by powerful data actors, but points towards careful ways forward at the practical intersections of environmental and data justice.

DISCUSSANT RESPONSE

Environmental data power towards environmental data justice – Eric Nost[12]

The brief reports assembled here give us a sophisticated sense of how environment and data intersect in the world today. These intersections are epistemological in the sense that how we know the environment is increasingly mediated through datafication, as well as material, in the sense that data infrastructures such as server farms and sensors are immersed in flows of energy (electricity) and matter (for example, rare earth metals). What is at stake is how we individually and collectively come to know our world and harm, care, or otherwise relate to it.

Each contributor makes their own unique addition, but the overall picture is clear: we cannot keep thinking about data in the ways we are used to. Data technologies are far from neutral tools, but sites of struggle over futures; they may be powerful, but they are not without their limitations; they are not just instruments of rationality, but derive much of their power from their appeal to our emotions; they do not float around as disembodied objects, but are situated in the world; far from immaterial, they help us sense through their physicality. Not virtual, they are 'more than real' (McLean, 2019).

My own work on data, society, and environment leads me to think about three elements of environmental data power. First, I want to be careful about binaries that divide the material and the epistemological. We know that knowledge is power, driving a datafication that has all sorts of interesting philosophical and political economic dimensions. Knowing nature from satellite imagery is undoubtedly different from knowing it through long-term field work or life experience (though *how* different and to what end is another question). Knowing nature through data that circulate quickly and widely is probably more amenable to capital accumulation than nature as known (or felt) through unwieldy individual accounts. My point is simply that we should not stop here, as these epistemologies imply physical outcomes on the landscape. For instance, when we turn to data technologies to tell us about future sea level rise and storm surges, we are led to outputs that reflect the conditions of that data – their scope, their producers' interests, and so on. In turn, these outputs reshape the world in very concrete ways by influencing which places see investments in climate adaptation (Nost and Colven, 2022). More broadly, Gabrys (2020) illustrates how datafied environments such as smart forests are increasingly thought of *as* technologies (for addressing environmental and climate change), leading to their material expansion across rural and urban spaces.

Second, it is worth making explicit what we mean by 'power' in environmental data power – where is it found and what are its limits? Do we focus on the dominant and seemingly unstoppable forces of extractive platform capitalism, or emphasize its glitches (Leszczynski, 2019)? What forms of resistance do we look for – the individual, the collective, the productive (Ettlinger, 2018; Thatcher and Dalton, 2021)? In the vein of glitches, Halkort reminds us of the importance of 'attending to the multiple gaps, blind spots, and misrecognitions' within data and data technologies. Data's emotional dimensions may not represent a glitch (Fratczak notes they are features not bugs), but attending to them too reflects a similar feminist understanding of power as multiple and embodied. Critics of datafication may take for granted – or at least not trouble – the idea that data registers as rational when lamenting how it reduces relations to numbers. This misses the other emotional registers that give data their power. As Nelson et al (2022) ask, 'How [do] digital technologies affect the everyday experiences, emotions, and sensations of socio-ecological relations?'

Third, we should wonder what environmental data power is for. At the moment, it is certainly wielded for extractivism, profit, and surveillance, but also for sustainability, community, and I would emphasize, justice. I would argue that the call for a coalition between data and environmental justice struggles leads us to questions at the heart of what colleagues have called environmental data justice (EDJ) (Dillon et al, 2017; Vera et al, 2019; Murphy, 2022). What is at stake in centring data that misrepresent people and places through bias or absences? What is possible and just to do with existing data, and what do empowering data infrastructures look like? Here, I would recall the data ethics refrain that 'data are people', which Thatcher and Dalton (2021) cleverly turn into a call for collective action towards sovereignty: 'Our data are us so make them ours.' In my mind, EDJ echoes this, but reminds us that we should think critically about the relations that constitute 'us' – not everyone is in the data or has the capacity to hold them. Environmental data power then means building the power to both rectify this, mitigating harm, while also de-centring the datafication processes that make it consequential in the first place.

Authors' response to discussant

Nost's points here are extremely welcome and pertinent – we are especially inspired by his insistence on maintaining the intricate entanglements between the epistemological and the material. In different ways, each contribution demonstrates that, in the context of environmental data especially, the material formations under discussion – whether technologies, infrastructures, environments, or institutions – are all co-constituted by the epistemological grounding and ideological contexts behind these different projects.

However, to brush against the grain slightly, the issue with centring beneficial, just, and 'ethical' data-driven applications, however much they can be directed towards justice (and especially in place of sustained critique), is that there is not a shortage of discussions surrounding ostensibly 'positive' uses of data technologies. Media, funding, and policy attention is extremely focused on ethical artificial intelligence (AI), technology for good, citizen science, and other mechanisms by which community power and engagement can be, at best, directed towards projects that benefit particular philanthropic interests and, at worst, co-opted towards more instrumental and profit-driven ends.

There is, however, a shortage of engaged critique of techno-solutionism and its embeddedness within environmental practices, especially in the public sphere. Putting more sustained and persistent critiques into the world, especially when aligned with practice that truly decentres these logics, is at this point just as important as identifying potential points to apply hope and form more just templates. Unfortunately, many of us as researchers are tied into funding structures and mechanisms which are built for such co-optation, as universities are by no means neutral actors in environmental data politics (for great, critical reflection on this double-bind, see Liboiron, 2021). By necessity, environmental justice organizations and researchers are usually stuck using certain kinds of technologies, infrastructures, and funding of green capitalists to do their work, and we need continual reflection on what this means for relationalities across research, data, and more-than-human ecologies. While we need to accept and acknowledge the inescapability of double-binds, critique may still be our most powerful weapon for identifying the points at which environmental data power may be seized towards a truly transformative environmental data justice, which will come, as Nost reiterates, by 'de-centring the datafication processes that make it consequential in the first place'.

Notes

[1] Discussant
[2] Facilitator
[3] University College Dublin

[4] Vrije Universiteit Brussel
[5] Vrije Universiteit Brussel
[6] See: https://science.nasa.gov/citizenscience
[7] See: https://cordis.europa.eu/project/id/101037648
[8] See: https://socio-bee.eu/?page_id=697
[9] See: https://ecsa.citizen-science.net/2016/05/17/10-principles-of-citizen-science/
[10] Lancaster University
[11] University of Sheffield
[12] University of Guelph, Canada

References

Ahmed, S. (2004) *The Cultural Politics of Emotions*. Edinburgh: Edinburgh University Press.

AQUAREAD (nd) Salinity Water Testing Sensors & Equipment. Available at: www.aquaread.com/sensors/salinity

Barad, K. (2012) On touching – the inhuman that therefore I am. *Differences*, 23(3): 206–223. https://doi.org/10.1215/10407391-1892943

Bonney, R., Phillips, T. B., Ballard, H. L., and Enck, J. W. (2015) Can citizen science enhance public understanding of science? *Public Understanding of Science*, 25(1): 2–16.

Borkin, M., Bylinskii, Z., Kim, N., Bainbridge, C., Yeh, C., Borkin, D., Pfister, H., and Oliva, A. (2015) Beyond memorability: visualization recognition and recall. *IEEE Transactions on Visualization and Computer Graphics*, 22.

Bowker, G. and Star, S. (1999) *Sorting Things Out: Classification and Its Consequences*. Cambridge, MA: MIT Press.

Bresnihan, P. and Brodie, P. (2021) New extractive frontiers in Ireland and the Moebius strip of wind/data. *Environment and Planning E: Nature and Space*, 4(4): 1645–1664.

Bresnihan, P. and Brodie, P. (2023) Data sinks, carbon services: waste, storage, and energy cultures on Ireland's peat bogs. *New Media and Society*, 25(2): 361–383.

Brighenti, A. M. (2018) The social life of measures: conceptualizing measure-value environments. *Theory, Culture & Society*, 35(1): 23–44.

Brodie, P. (2020a) Climate extraction and supply chains of data. *Media, Culture and Society*, 42(7–8): 1095–1114.

Brodie, P. (2020b) 'Stuck in Mud in the Fields of Athenry': Apple, Territory, and Popular Politics. *Culture Machine*, 19. Available at: https://culture machine.net/vol-19-media-populism/stuck-in-mud-in-the-fields-of-athe nry-patrick-brodie/

Brodie, P. (2021) Hosting cultures: placing the global data centre 'industry'. *Canadian Journal of Communication*, 48(4): 151–176.

Bucher, T. (2018) *If... Then: Algorithmic Power and Politics*. New York: Oxford University Press.

Butchart, G. C. (2015) Touch, communication, community: Jean-Luc Nancy's semiotic phenomenology. *Metodo: International Studies in Phenomenology and Philosophy*, 3(1): 221–244.

Carswell, S. (2022) 'Data centres now consuming more electricity than rural homes – CSO'. *The Irish Times*, 3 May. Available at: www.irishtimes.com/news/ireland/irish-news/data-centres-now-consuming-more-electricity-than-rural-homes-cso-1.4868221

Chakrabarty, D. (2021) *The Climate of History in a Planetary Age*. Chicago and London: University of Chicago Press.

Climate Outreach (2021) 'Our theory of change [online]'. Available at: https://climateoutreach.org/about-us/theory-of-change/

Deckard, S. (2016) World ecology and Ireland: the neoliberal ecological regime. *Journal of World Systems Research*, 22(1): 145–176.

Dickinson, J. L. and Bonney, R. E. (2012) *Citizen Science: Public Participation in Environmental Research*. New York: Cornell University Press.

D'Ignazio, C. and Bhargava, R. (2020) 'Data Visualization Literacy: A feminist starting point'. In: M. Engebretsen and H. Kennedy (eds) *Data Visualization in Society*. Amsterdam: Amsterdam University Press, pp 207–222.

Dillon, L., Walker, D., Shapiro, N., Underhill, V., Martenyi, M., Wylie, S., et al (2017) Environmental data justice and the Trump administration: reflections from the Environmental Data and Governance Initiative. *Environmental Justice*, 10(6). https://doi.org/10.1089/env.2017.0020

Ettlinger, N. (2018) Algorithmic affordances for productive resistance. *Big Data & Society*, 5(1): 205395171877139. https://doi.org/10.1177/2053951718771399

Gabrys, J. (2014) Programming environments: environmentality and citizen sensing in the smart city. *Environment and Planning D: Society and Space*, 32(1): 30–48.

Gabrys, J. (2019) Sensors and sensing practices: reworking experience across entities, environments, and technologies. *Science, Technology, & Human Values*, 44(5). https://doi.org/10.1177/0162243919860211

Gabrys, J. (2020) Smart forests and data practices: from the Internet of Trees to planetary governance. Big Data & Society, 7(1): 205395172090487. https://doi.org/10.1177/2053951720904871

Goldstein, J. and Nost, E. (eds) (2022) *The Nature of Data: Infrastructures, Environments, Politics*. Lincoln, NE: University of Nebraska Press.

Gray, J. (2020) 'The Data Epic: Visualization practices for narrating life and death at a distance'. In: M. Engebretsen and H. Kennedy (eds) *Data Visualization in Society*. Amsterdam: Amsterdam University Press: 313–328.

Groot, B. and Abma, T. (2022) Ethics framework for citizen science and public and patient participation in research. *BMC Medical Ethics*, 23(1).

Halkort, M. (2021) 'Post-human Tactility: Sensing death in the Mediterranean Sea'. In: D. Zyman and M. Reyman (eds) *Oceans Rising*. Berlin: Sternberg Press, pp 42–45.

Halkort, M. (2022) 'The Necropolitics of Innovation'. in R. Hoyng and G. P. Lei Ching (eds) *Communication Innovation and Infrastructure: A Critique of the New in a Multipolar World*. East Lansing, MI: Michigan State University Press, pp 167–186.

Haraway, D. (2007) When Species Meet. Minneapolis: Minnesota University Press.

Harold, J., Lorenzoni, I., Shipley, T., and Coventry, K. (2020) Communication of IPCC visuals: IPCC authors' views and assessments of visual complexity. *Climatic Change*, 158: 255–270.

Haroz, S. and Whitney, D. (2012) How capacity limits of attention influence information visualization effectiveness. *IEEE Transactions on Visualization and Computer Graphics*, 18: 2402–2410.

Haroz, S., Kosara, R., and Franconeri, S. (2015) ISOTYPE visualization: working memory, performance, and engagement with pictographs. *Proceedings of the 33rd Annual CHI Conference on Human Factors in Computing Systems*, pp 1191–1200.

Hogan, M. (2015) Data flows and water woes: The Utah Data Center. *Big Data and Society*, 2(2). https://doi.org/10.1177/2053951715592429

Hogan, M. (2018) Big data ecologies. *Ephemera: Theory and Politics in Organization*, 18(3): 631–657.

Irwin, A. (1995) *Citizen Science: A Study of People, Expertise, and Sustainable Development (Environment and Society)*. London: Routledge.

Kennedy, H. and Hill, R. (2018) The feeling of numbers: emotions in everyday engagements with data and their visualization. *Sociology*, 52(4): 830–848.

La Puig de Bellacasa, M. (2017) Matters of care: speculative ethics in more than human worlds. *Posthumanities*, 41. Minneapolis: University of Minnesota Press.

Lally, D. (2019) 'The sacred fire of a data centre'. *Strelka Magazine*, 10 October. Available at: www.academia.edu/41096301/The_Sacred_fire_of_the_Data_Center_Strelka_Donal_Lally

Lehuedé, S. (2022) Territories of data: ontological divergences in the growth of data infrastructure. *Tapuya: Latin American Science, Technology and Society*, 5(1).

Leiserowitz, A. (2006) Climate change risk perception and policy preferences: the role of affect, imagery, and values. *Climatic Change*, 77: 45–72.

Leszczynski, A. (2019) Glitchy vignettes of platform urbanism. *Environment and Planning D: Society and Space*, 38(2): 189–208. https://doi.org/10.1177/0263775819878721

Liboiron, M. (2021) *Pollution Is Colonialism*. Durham, NC: Duke University Press.

McLean, J. (2019) *Changing Digital Geographies: Technologies, Environments and People* (1st edition). Cham: Palgrave Macmillan.

Metag, J., Schäfer, M. S., Füchslin, T., Barsuhn, T., and Kleinen-von Königslöw, K. (2016) Perceptions of climate change imagery: evoked salience and self-efficacy in Germany, Switzerland, and Austria. *Science Communication*, 38(2): 197–227.

Moss, S. (2022) '21 Irish data centers planned outside greater Dublin, after city ban'. Data Centre Dynamics, 19 December. Available at: www.dat acenterdynamics.com/en/news/21-irish-data-centers-planned-outside-greater-dublin-after-city-ban/

Murphy, M. (2022) Chemical futures and environmental data justice. *UnderCurrents: Journal of Critical Environmental Studies*, 21: 45–48. https://doi.org/10.25071/2292-4736/40349

Nærland, T. U. (2020) 'The Political Significance of Data Visualization: Four key perspectives'. In: M. Engebretsen and H. Kennedy (eds) *Data Visualization in Society*. Amsterdam: Amsterdam University Press, pp 40–47.

NASA (2009) With a pinch of salt. Available at: https://climate.nasa.gov/news/58/with-a-pinch-of-salt/

Nelson, I. L., Hawkins, R., and Govia, L. (2022) Feminist digital natures. *Environment and Planning E: Nature and Space*: 251484862211231. https://doi.org/10.1177/25148486221123136

Nierse, C. (2019) *Collaborative User Involvement in Health Research Agenda Setting*. Amsterdam: Vrije Universiteit.

Nost, E. and Colven, E. (2022) Earth for AI: a political ecology of data-driven climate initiatives. *Geoforum*, 130: 23–34. https://doi.org/10.1016/j.geoforum.2022.01.016

O'Brien, C. (2022) 'Microsoft signs power purchase deals to support Ireland's renewable energy infrastructure'. *The Irish Times*, 23 November. Available at: www.irishtimes.com/business/2022/11/23/microsoft-renewable-contra cts-to-contribute-almost-30-of-power-purchase-target/?

O'Neill, S. (2020) More than meets the eye: a longitudinal analysis of climate change imagery in the print media. *Climatic Change*, 163: 9–26.

O'Neill, S., Boykoff, M., Day, S., and Niemeyer, S. (2013) On the use of imagery for climate change engagement. *Global Environmental Change*, 23: 413–421.

Paananen, J. and Nasr, E. (2021) 'Grid-interactive data centers: enabling decarbonization and system stability'. Eaton. Available at: www.eaton. com/content/dam/eaton/markets/data-center/eaton-microsoft-grid-inte ractive-whitepaper-wp153031en.pdf

Roach, J. (2022) 'Microsoft datacenter batteries to support growth of renewables on the power grid'. Microsoft Source, 7 July. Available at: https://news.microsoft.com/source/features/sustainability/ireland-wind-farm-datacenter-ups/

Swinhoe, D. (2023) '11 data centers in Dublin set to rely on Ireland's gas network for power'. Data Centre Dynamics, 24 April. Available at: www.datacenterdynamics.com/en/news/11-data-centers-in-dublin-set-to-rely-on-irelands-gas-network-for-power/

Taffel, S. (2021) Data and oil: metaphor, materiality and metabolic rifts. *New Media and Society*, 25(5): 980–998. https://doi.org/10.1177/146144 48211017887

Thatcher, J. E. and Dalton, C. M. (2021) *Data Power: Radical Geographies of Control and Resistance*. London: Pluto Press.

Velkova, J. (2021) Thermopolitics of data: cloud infrastructures and energy futures. *Cultural Studies*, 35(4–5): 663–683.

Vera, L. A., Walker, D., Murphy, M., Mansfield, B., Siad, L. M., Ogden, J., and EDGI (2019) When data justice and environmental justice meet: formulating a response to extractive logic through environmental data justice. *Information, Communication & Society*, 22(7): 1012–1028. https://doi.org/10.1080/1369118X.2019.1596293

Wahl-Jorgensen, K. (2018) *Emotions, Media and Politics*. Cambridge: Polity Press.

Whitington, J. (2016) Carbon as a metric of the human. *PoLAR: Political and Legal Anthropology Review*, 39(1): 46–63.

8

Data and Technological
Spatial Politics

Yaya Baumann,[1] Janna Frenzel,[2] Emanuel Guay,[3] Leonora Indira King,[4] Alex Megelas,[5] Alessandra Renzi,[6] Julia Rone,[7] Sepideh Shahamati,[8] Hunter Vaughan,[9] Tamara Vukov,[10] Rob Kitchin,[11] and Jo Bates[12]

Introduction

This chapter traces the contested politics of data, shifting scales from the transnational to the national and local levels: from questions of the (extra-) territoriality of data, the role of jurisdictions and contested 'technical territories' (Munn, 2023a) to the concrete lived spaces where data are produced, stored, and circulated. The different contributions thus zoom in from global geopolitical struggles over digital sovereignty and hegemony over data infrastructure, to local contestations over subsea cable networks and landing stations, data centres, as well as neighbourhood gentrification driven by the development of artificial intelligence (AI). This multiscalar approach to data politics aims to emphasize the tensions between the abstract global logics of data circulation and the local realities of data, between historical state and corporate projects of extending data territories as a form of 'domination'. It points to the localized effects of such projects, including gentrification, expropriation, and the colonial erasure of local knowledges and sovereignty.

To deal with these complex issues, we introduce three key types of data power that help us understand better the agency not just of states, corporations, and citizens but also of data infrastructures themselves as mediating between different spaces, territories, jurisdictions, and temporalities. The three key concepts we introduce are:

- *Hegemonic data power* – to describe the ways in which states and corporations are mobilizing (and in some cases weaponizing) data infrastructure to further their geopolitical ambitions. The concept of hegemonic data power at the same time invites questioning of the notion of hegemony in our multi-polar world and opens a broad field for exploring contested state–state relations (as observed in the recent rise of the doctrine of digital sovereignty) but also contested state–corporate relations (as seen in the recent banning of tik tok in the US). Recent decades have given rise to tech corporate monopolies that have become key players on the global stage, competing with states and taking on some state functions, but also collaborating with and furthering state ambitions, depending on changing configurations of power. Hegemonic data power has a long history in colonial ambitions (see Vaughan's contribution) and is currently undergoing a process of reconfiguration (see Rone's section).

- *Powering data* – the concept of powering data points to the environmental consequences of contested data power and draws attention to the materiality of data infrastructures, their energy and water demands, and the ways in which they are imbricated in processes of extraction at the local level – be it of minerals (Nobrega and Varon, 2020), electricity, or water. In fact, the reconfigurations of hegemonic data power and the race to develop AI can hardly be understood without paying attention to how states and corporations collaborate and compete over the extraction of resources at a planetary scale. At the same time, numerous researchers have put forward ideas for alternatives to the current AI models and extractivist approaches to data (see Frenzel and Renzi's contribution).

- *Data counter-power* – this concept draws attention to the multiple fields of resistance to both hegemonic data power and extractivist modes of powering data. Across the Global South and the Global North there have been bottom-up movements opposing data centre construction on the grounds of both their environmental consequences and the lack of democratic participation in decision-making (for example, Rone, 2023; Brodie, 2023). Data counter-power is also at the basis of nascent ideas around grassroots green AI and anti-extractivist data projects, data literacy education, as well as more established data privacy protection efforts and surveillance self-defence. In a similar vein, we frame approaches to data sovereignty as contributing to the emergence of data counter-power formations at a translocal scale. As a form of counter-power, data sovereignty emerges as a discourse on technological and cultural autonomy, as well as a set of practices developed by tech collectives of/for underrepresented or marginalized groups to assert their presence and agency (Couture and Toupin, 2019). This is visible in the work of indigenous scholars and practitioners who advocate for Indigenous Data Sovereignty as a framework to control data, its infrastructure, and use (Bruhn, 2014; Kukutai and Taylor, 2016), as well as indigenous epistemologies to produce and mobilize data (Renzi and Radjawali, forthcoming; Foxworth and Ellenwood, 2023). At the same time, data infrastructure is envisioned as a pivotal element in building data sovereignty. This includes both indigenous data and research infrastructure (McMahon et al, 2015; Duarte, 2017; Carroll et al, 2020), and philosophical considerations and experiments in developing AI that is integrated in indigenous ontologies (Lewis et al, 2018). For grassroots and social movements, visions for data sovereignty include projects around data commons, free/open source software, autonomous servers, and encryption (Haché, 2017).

At the micro level, several of the contributors to this volume explore community activism through a case study of Montreal, where local activists oppose processes of gentrification and displacement driven by an emerging AI ecosystem meant to boost Canada's innovation and platform economies. We home in on instances of community mapping that produce data in a fair and equitable way; data that empowers communities to resist gentrification and expropriation and to support situated knowledges.

The idea of data counter-power driving socio-political formations that are (1) spatially situated, (2) often networked, and (3) constantly reframing how data can be prod/used for justice, draws attention to how data themselves are implicated in multiscalar struggles to wield power. Such examples fall in a long line of data activism projects, including attempts to fight corruption through data analysis and crowd-sourced data (Mattoni and Odilla, 2021). In examples such as these, data become not simply stakes, that is issues and objects of political struggle, but also repertoires, that is tools of political struggle (Beraldo and Milan, 2019).

In what follows, we lay out some of these multiscalar processes and tensions, starting with how settler and extractive colonialism are constantly reshaping physical and political geographies of power through amphibious infrastructural projects. Our look at the geopolitical dimension of data power(s) continues with a genealogy of hegemonic forms of data sovereignty and with some considerations about the environmental impact of data and AI economic innovation. We then show how forms of counter-power emerge and traverse these geographies as they harness social movement practices, research advocacy, and knowledge/data commons.

Spatial data colonialism and amphibious infrastructural data power – Hunter Vaughan

Coastlines are liminal spaces of interaction and points of contest between local and international zones of policy and law. They are spaces of shifting sovereignty and invasion, with local agency being challenged historically by military colonization, and today by the data infrastructural colonialism of tech insurgence. As climate destabilization threatens coastal spaces with increasing sea level rise, extreme weather events, and species extinction, coastlines will prove a crucible for the challenges of community agency and environmental preservation amid accelerating climate crisis and exponential digital growth.

Coastlines also are unique in shaping – and being shaped by – oceanic amphibious infrastructures. Telecommunications infrastructural growth has been integral to colonial expansion, capitalist cartography, and the great acceleration of anthropogenic climate change from the 1850s to today. Since the first subsea telegraph cables were laid, cable landing stations and networks have been deeply enmeshed in the routes of maritime power and economics and the dynamics of resource use. Subsea cables are now central to the emergence of newly consolidated configurations of big tech power that challenge traditional power structures of the nation state and are also operative in the emergence of alternative energy futures.

The socio-technical morphology of infrastructures that have intruded and implanted on their shores carries historical tensions between traditional local knowledge and imposed hegemonic epistemologies – and now, increasingly, the guidance of algorithmic decision-making. New infrastructural configurations also build upon the role that coastal infrastructures of communication and energy have played in the post-colonial process of local economic development, resource subsistence, and identity construction, and as such present a key battlefield for agency, decision-making, and heritage preservation in the face of rapidly expanding digital hegemony. Consequently, such configurations have featured in recent work on digital media infrastructure studies (Johnson, 2023), energy futures (Watts, 2019; de Onís, 2021), and the geopolitics of colonialism and extraction (Ruiz, 2021).

Scholarly coverage of subsea cable networks and landing stations is beginning to emerge (Pasek et al, 2023; Vaughan et al, 2023), especially as these largely invisible infrastructures become more revealed through studies of their relevance to national security (Bueger et al, 2021; Keller, 2023) and marine ecology. Subsea networks have played a crucial role in the historical global organization of geopolitical power, technological expansion, and social connectivity. Today, they are one of the Internet's most vital infrastructures, carrying almost 100 per cent of transoceanic data traffic between continents (Starosielski, 2015). Geopolitically, they are bound up in the shift to a post-colonial and neo-imperial era, where previous global superpowers are being replaced by post-political forces of big tech and energy companies (Harvey, 2001; Kwet, 2019; Szeman, 2019). Despite growing literature on the geography and sovereignty of data (Bratton, 2015; Duarte, 2017; Mueller, 2017; Graham and Dittus, 2022), and the environmental impacts of the Internet (Maxwell and Miller, 2012; Hogan, 2015; Malmodin and Lundén, 2018; Marks et al, 2020), few studies have examined the role of subsea networks. To do so, we must also acknowledge the crucial role these infrastructures played in the shift from maritime trade to submarine communication as dominant modes of transoceanic empire building, colonial occupation, and extraction (the latter now largely being in the form of data).

Coastlines were greatly altered, and communities built around the physical intrusions and economic systems of these cable networks. In many cases, unlike the record of social impacts and local grassroots resistance to data centre construction (Brodie, 2020; Lehuedé, 2022; Rone, 2023), these have yet to manifest in clear acts of resistance. This genealogy is marked instead by legacies of percolating political tension, social inequality, and socioecological change that may not be easily attributable to current infrastructures, but provide deeply contested backdrops for the arrival of digital tech empire building. For example, the west shore of the island of Oʻahu, where three major transpacific cable landing stations exist, has been a site of conflict and tension with local communities for several decades. This economically disenfranchised area is home to many Native Hawaiians, with some groups of people living houseless on the beach, and a crucial site of protest for Hawaiian sovereignty. The local community has struggled against lack of economic opportunities and negative portrayals in news media.

While the cable stations and their adjacent networks, crucial to transpacific connectivity of island nations and territories as well as US connectivity with Oceania and East Asia, may not have caused the disenfranchisement of Native communities, their presence is an infrastructural reminder of the global inequalities and invasive extractive dynamics by which the digital divide extends legacies of industrial military colonialism. Local communities are excluded from the economic and communications benefits of these infrastructures, while negative representations of this community are disseminated through the very infrastructures that have been built upon their unacknowledged sovereign land.

More directly, along the transatlantic route that remains the most highly trafficked data route in the world, telecommunications development is inseparable from imperial history, in particular England's 19th- and 20th-century colonial military expansion (Winseck and Pike, 2007; Campling and Colás, 2021). The subsea telecommunications implications of British colonial history are deeply evident throughout its transatlantic cartography, including: Bermuda, which has been a British colony since 1609 and remains a British Overseas Territory (Stockwell, 2008); Nova Scotia, which offered the crux of English–French colonial struggle over Canada from the mid-1800s until well into the 20th century (Conrad, 2022); and Ireland, which was a British colony for many years until its independence in 1921 (Kee, 2000). In rural west Ireland, from Valentia to Galway to Killala, debates around external investment into locally planned data infrastructures are loaded with vestiges of foreign occupation and complex national politics between urban Dublin and the rural west and further complications regarding nationalized strategies and the mandates of Ireland's EU involvement.

From Puerto Rico to Hong Kong to Fortaleza, to the entire African continent being targeted for the 2Africa network financed by China Mobile, Meta, and Orange, subsea cable networks both carry and symbolize key shifts in 21st-century data power, whether in terms of geopolitics, resource, labour, or information. They are an icon and index of the invisibility of this circulation, converting liminal geographical and geological spaces into points of transformation between different kinds of energy, value, and capital. Moreover, they connect the constellation of terrestrial networks – with all their interlocking and shifting dynamics of power – from country to country, continent to continent.

Digital sovereignty and the geopolitical struggle over data and technological spatial politics – Julia Rone

While data infrastructures such as telegraph cables, data centres, or subsea cables have a very local presence, they have always also been imbricated in projects of hegemonic global data power (Starosielski, 2015) led by empires and/or corporations entering into complex relations of collaboration and competition with each other (Winseck, 2019). These processes have not gone uncontested and protests have been common (Brodie, 2020; Rone, 2023). It is against this background of global battles for hegemony in data power, that many local conflicts around data infrastructure unfold, tackling issues such as gentrification, environmental deterioration, and lack of democratic participation. In this section, I trace some of the key aspects of the global geopolitical struggle for data infrastructural dominance, focusing on key developments over the last 30 years, including the rise of the notion of 'digital sovereignty' and the ways in which data infrastructures have challenged and enriched traditional notions of territory and spatiality.

Data sovereignty is not always related to forms of counter-power. Most accounts of the rise of the concept of digital sovereignty date it back to the 2010s, when authoritarian states such as China and Russia first developed their doctrines of digital sovereignty, soon to be followed by democratic states, reacting to the 2013 Snowden revelations and the 2016 Cambridge Analytica scandal. Nevertheless, such debates are much older and can be traced to the rise of the New World Information and Communication Order movement (Nordenstreng and Schiller, 1978), which aimed to make global media communications more equitable. The problem of state sovereignty vis-à-vis powerful global corporations such as IBM was a central issue already in the bestselling Nora and Minc (1980) report to the president of France on the computerization of society. What lies between these early discussions of state sovereignty over communications and the so-called 'return of the state' in the 2010s (Haggart et al, 2021) is the rise of the Internet, which started as a US-military project developed in the late 1960s (the ARPANET), underwent privatization in the 1990s and saw the consolidation of several US big tech monopolies over the 2000s, including the world's biggest companies in terms of market capitalization: Alphabet (parent company of Google), Apple, Meta (parent company of Facebook), and Microsoft.

The period of expansion of Internet's penetration and user base (as well as its increasing privatization) coincided with a hegemonic vision of an open global Internet as a guarantee for freedom and democracy. This powerful socio-technical imaginary of the open Internet was consistently promoted by the US and served well its political economy and geopolitical interests in a period in which most data infrastructure, hardware, and software were dominated by US companies (Powers and Jablonski, 2015). It has been argued that the spatial expansion of US private companies and government projects led to a 'weaponization of infrastructure' in which the US government was able to both *pull* information from data infrastructure through surveillance programmes (such as the ones revealed by Edward Snowden) as well as to *push* information into foreign societies by supporting different tools for circumventing government censorship (Ortiz Freuler, 2022).

Nevertheless, the hegemony of the US government and US big tech companies has been recently challenged by the rise of East Asian governments and corporations alike in the field of subsea cable construction (Winseck, 2019) but also in cloud infrastructure (Munn, 2023b) and especially in the development of 5G standards, where China has established itself as a global leader (ten Oever, 2022). In the late 2010s, the European Union also developed a doctrine of digital sovereignty as a way to foster innovation and more control over its citizens' data. In the context of rising trade wars between the US and China and the disruption of supply chains during the COVID-19 pandemic, the EU has attempted to assert itself as a global geopolitical player beyond its usual normative soft-power, through an ambitious combination of new regulations – such as the AI Act, as well as industrial strategy (Pohle and Thiel, 2020; Baur, 2023; Broeders et al, 2023).

The shrinking of the percentage of US Internet users as compared to global numbers and the rise of other global digital players such as China, and to a lesser extent the EU, have led first of all to a re-networking of the global Internet, which has become increasingly fragmented, a so-called 'splinternet'. In this new situation, the US government has used legislation to attempt to *cut* particular countries and their users from global supply chains and platform services (Ortiz Freuler, 2022). Similar actions have been taken also by the EU – for example, with its introduction of screening for foreign direct investment – a tool very clearly targeting Chinese investment in 5G projects in Europe (Broeders et al, 2023).

Second, the rising importance of markets other than the US has complicated the relationship and balance of power between profit-pursuing private US corporations and the US government. On the one hand, we have seen tech corporations increasingly taking on state functions such as attempting to develop their own currencies or hosting highly sensitive government data (Utrata, 2022). On the other hand, US corporations have increasingly accepted to 'play by the rules' of other states. For instance, both Microsoft and Google have embraced European notions of 'digital sovereignty', promising to localize data in Europe (Fox-Martin, 2022; Sanders, 2022). Of course, the extent to which US corporations have indeed escaped the reach of the US government is questionable. A decision by the French government to host national health data of French citizens on Microsoft Azure has led to a strong contestation by civil society and the national data protection authority on the grounds that the US government could still potentially ask for access to information under the US Cloud Act (Pannier, 2021).

All in all, rather than talking about hegemonic data power as the power of a single hegemon (formerly the US), what we see currently is a competition between different hegemons – both states and private corporations – that plays out through the development of strategic technology but also through careful placement of data infrastructure at key domestic locations as well as global chokepoints (such as Hong Kong or Singapore) where East and West meet (Munn, 2023b). In this 'war of position' we often see collaborative projects between competitors, with Chinese and US state companies sometimes collaborating in building subsea cables, for example (Winseck, 2019; Munn, 2023b).

Ultimately, the geopolitical struggle over data and technological spatial politics takes place through re-territorialization of data as seen in attempts by China and the EU to enhance data localization and guarantee that the data of their citizens will be processed on their territory. But it also takes place through the expansion of data infrastructure across the globe that allows each of the big geopolitical players, but especially the US and China, to *pull* data from and *push* information into other territories. A case in point was the massive data hack of the African Union building in Addis Ababa, Ethiopia, in which China was deeply implicated. The hack involved not only the bugging of the building but also the transfer of data between midnight and 2 a.m. from the Union's servers to Shanghai, 8,000 km away (Munn, 2023b).

It is this important affordance of data infrastructure to be able to operate both within and beyond any given jurisdiction that has led to the introduction of the concept of 'technical territories' (Munn, 2023b). Technical territories are both highly localized *and* extend beyond the local to far away territories in their operations. As such, technical territories become key loci of power – loci where both state and corporate power is performed, but also where it is increasingly contested – by states and citizens alike as we will see in the next sections of this chapter.

Powering the data ecosystem: AI and climate change – Janna Frenzel and Alessandra Renzi

One such example of technical territory is located in Canada, where, in recent years, the city of Montreal has become a hub for AI research and development – or, as the industry has popularized it, an 'AI ecosystem'.[13] This loose network of entities includes 5,000 businesses and start-ups with nearly 100,000 workers, academic R&D units with special appointees, and some 9,000 students (Brandusescu, 2021). Important players are the University of Montreal, Element AI, the Quebec Artificial Intelligence Institute (known as MILA), but also multinational tech giants such as Google, Meta, and Microsoft (Startup Montréal, 2021).[14]

The ways in which the AI industry's underlying infrastructures and their material manifestations are changing the urban and environmental fabric of Montreal can be described as a specific instance of what Mimi Onuoha (2018) calls 'algorithmic violence'. But here, algorithmic violence is not limited to the impacts of AI data and software. Instead, it includes a variety of layers that obfuscate different kinds of violence. One of these layers pertains to the digital infrastructure for climate change mitigation that heavily relies on greenwashing and resource extraction. Another layer covers how the AI industry's power to purchase or rent real estate for office spaces and R&D units, and the arrival of white-collar tech workers and students contribute to an ongoing housing crisis. As the next sections will explain more in detail, this kind of spatial power reconfigures access to basic needs such as housing within primarily low-income neighbourhoods.[15]

The identity of many actors within the Montreal 'AI ecosystem' is firmly grounded in the 'AI for good' or 'responsible AI' narrative. From its beginnings, the availability and affordability of hydropower has played a key role in the marketing of Montreal's AI ecosystem as 'greener' than other locations. For instance, at the 2022 AI World Summit in the Americas held in Montreal, representatives from Forum AI and Investment Québec praised Montreal's 'unique' AI ecosystem as 'green from the start' (and allegedly contributing to 'saving the world').

While statements like this can easily be dismissed as just another iteration of greenwashing for marketing purposes, they also conceal the long history of the links between extractive infrastructure development for hydroelectricity, Quebec nation-building, and indigenous dispossession in the province (see Desbiens, 2013; Nungak, 2017). The question of environmental impacts of AI systems cannot be limited to assessing an AI system in isolation – it needs to include the question of what purposes these systems fulfil, who benefits and who loses out, and whether the systems in question are desirable. Reducing the question of AI's environmental impact to whether or not it is powered by renewable energy not only ignores these complex connections; it also obscures the fact that AI systems are employed by the fossil fuel industry to identify yet unextracted reservoirs and distribute, refine, and market oil and gas (Donaghy et al, 2020).

In techno-optimist narratives, AI and machine learning are often touted as a 'solution' to climate breakdown. For instance, the agenda of 'Sustainability for the Digital Age' (SDA) (which involves Mila and CIFAR, among others), proposes that 'digital disruptors' can act as levers for systemic change, leading to a 'climate-safe and equitable world' (SDA, 2020).

While it can be a useful tool in climate monitoring and modelling, AI also contributes to resource extraction and greenhouse gas emissions. AI is built on specific sets of knowledge in the form of data, but also relies on raw materials for hardware production, energy, labour, affect, and networked infrastructures (Pasquinelli and Joler, 2020; Crawford, 2021). As AI models grow continuously larger, for example, most recently with the development of generative AI, the more computationally intensive these models become to train and deploy, which in turn means they consume more electricity and other computing-related resources (Amodei and Hernandez, 2018; Schwartz et al, 2019; Saenko, 2020; de Vries, 2023; Li et al, 2023; Zewe, 2023).

However, measuring the impacts of AI systems is not an easy task since there is currently a lack of reporting mechanisms that would be necessary for a systematic assessment. In the absence of any meaningful regulation, it is up to the developers of AI systems to determine what environmental factors they want to consider. The lack of data and accountability mechanisms for the material footprint of digital technologies in general, and of AI in particular, means that they are also mostly absent from institutional pledges to sustainability. The University of Montreal, for instance, did not include computing infrastructure and data processing in its overarching sustainability strategy for 2021–2023. Housing was not included either (Université de Montréal, 2021).

Could AI be done differently? Possible taxonomies and computation principles that would better respect planetary boundaries have been put forward by scholars and practitioners, such as sustainable computing (Toyama, 2015), limits-aware computing (Chen, 2016), computing sufficiency and self-sufficiency (Hilty, 2015), benign computing (Raghavan, 2015), and degrowth computing (Sutherland, 2022). 'Tiny AI' is gaining traction in industry (Hao, 2020). It remains to be seen whether any of these ideas will find widespread adoption in the field of AI.

While extractive conditions remain in place throughout the AI supply chain and 'externalities' – be it environmental impacts of hardware production or the displacement of urban residents through gentrification – are largely ignored, the hard questions around the AI industry's power, ownership and control, financial gain, and the extraction of liveable spaces on both the local and planetary scales must be asked. The 'ecosystem' metaphor may signal the arrival of a specific realm of algorithmic violence where boundaries between business, academia, and government are increasingly blurred, and where points of intervention are not easily identifiable. As the next sections discuss, this may limit the potential for resistance, unless intersectional organizing succeeds at making alliances that are able to look at the whole picture of algorithmic oppression.

Tech-led gentrification and community hardship: the case of Parc-Extension, Montreal – Yaya Baumann and Emanuel Guay

Beyond the greenwashed façade of AI and knowledge industries, their impacts on gentrification is particularly salient in Parc-Extension (Parc-Ex), a low-income neighbourhood in Montreal which is home to a large and diverse community of first- and second-generation immigrants, as well as numerous support networks and community services. While processes of gentrification and displacement, along with the in-migration of predominantly white, better-off population and creative workers in Parc-Ex were previously sporadic trends, the opening of the Université de Montréal's new campus, the MIL, in 2019, and the creation of an AI supercluster on its outskirts signal a pivotal shift in the neighbourhood's socio-economic trajectory. Although this is new to Parc-Ex, the requalification of the area into a new hi-tech hub is paradigmatic of Montreal's strategy of targeting new spaces for accumulation. This strategy aligns with the city's post-industrial approach of encouraging the development of a creative knowledge economy to attract both local and international investments (Sprague and Rantisi, 2019). For instance, the construction of the MIL campus, whose name stands for 'Middle of the Island' and 'Montréal Innovative Lab', benefited from $350 million offered by the three levels of government (that is, municipal, provincial, and federal), while the AI supercluster also benefited from massive public investments (for example, $500 million in Foreign Direct Investments, and $230 million from the federal government).

In the midst of a persisting housing crisis, the city of Montreal's enthusiasm for this new hi-tech hub raises particular concerns, as its development has deep ties with the increased precarity of Parc-Ex's most marginalized communities. For instance, it has been shown that investments in these two projects have led to a sharp surge in average rent prices over the past few years, exacerbating the already precarious situation of many tenants, and putting a disproportionate pressure on the neighbourhood's rental market (Jolivet et al, 2022; Renzi et al, 2022). In turn, the support networks that have been vital to Parc-Ex's tenants in mitigating the adverse effects of gentrification, racism, exploitative working conditions, and social exclusion have been seriously shaken by this increased pressure. Furthermore, an increasing number of speculating landlords are acquiring properties, adopting an investment strategy that relies on evicting long-term tenants to maximize rent and attract a supposedly more 'profitable' clientele (Guay and Megelas, 2021). More alarming still, are the shared failure of both public and private actors to mitigate their negative impacts on the neighbourhood, let alone acknowledge their responsibilities.

Among the many consequences of tech-led gentrification in Parc-Ex, our sustained engagement with the community has led us to focus on four particularly pressing ones. A first major consequence is the increased residential precarity of the neighbourhood's most marginalized tenants. This can be seen in the significant increase in evictions since the opening of the MIL campus (Parc-Extension Anti-Eviction Mapping Project, 2020), and the new branding for Parc-Ex's rental market, which manifests itself in ads and discourses held by real estate developers. This new branding targets students as well as young professionals, and seeks not only to increase rent, but also to install (new) practices of gentrifiers (Jolivet et al, 2022), creating overlapping processes of gentrification, youthification, and studentification (Moos et al, 2019). This worsened residential precarity for long-term tenants comes with a set of challenges (for example, harassment, fear of displacement, increased housing costs burden), as well as increased difficulties for collective organizing, which is caused, among other things, by the displacement of tenants and community groups (Reese et al, 2010: 321). Notwithstanding this looming threat of displacement faced by community groups in Parc-Ex, the transformation of the neighbourhood, coupled with the disproportionate allocation of resources towards the tech industry, exacerbates the sense of powerlessness among community workers. They contend with mounting demands for support without a corresponding increase in available resources. A third consequence is that the development of community housing projects is becoming harder in Parc-Ex, as the cost of land goes up and real estate speculation increases competition to acquire buildings. Together with funding cuts for community housing projects, this notably means that community groups struggle with acquisition costs, are more likely to lose sites to private real estate developers, and have a harder time convincing public authorities to acquire sites and buildings in the neighbourhood (Guay and Drago, 2023). Finally, a fourth consequence is the reinforcement of a cynical and distrustful point of view on public institutions by the neighbourhood's most marginalized tenants, who perceive these institutions as inattentive to their needs and unable to support them when faced with a major crisis such as the potential or actual loss of their dwelling. This cynicism and distrust encourages disengagement and impedes the capacity of local community groups to lead campaigns for housing justice (Exentus, 2022).

The Parc-Ex Anti-Eviction Mapping Project – Sepideh Shahamati and Tamara Vukov

The Parc-Ex Anti-Eviction Mapping Project (PEAMP)[16] emerged in 2019 to highlight housing struggles and to support residents and community groups of Parc-Ex in their fight against intensifying gentrification and displacement in the neighbourhood (Parc-Extension Anti-Eviction Mapping Project, 2020). By using digital mapping tools, quantitative and qualitative data, research, writing, and creative practices, the project works to highlight and counter the role of institutions and businesses such as the University of Montreal and tech companies in driving housing speculation, rent increases, and displacement in the Parc-Ex neighbourhood.

In resonance with Maharawal and McElroy (2018), this project adopts a counter-mapping approach to make the landscapes of displacement and resistance visible. As described by Dalton and Stallmann (2018), counter-mapping, as a combination of critical tools and practices for social change, can create opportunities 'for grassroot data science initiatives' (Dalton and Stallmann, 2018: 93). Through this counter-mapping approach, we do not abandon the use of maps for their long traits of manipulation, distortion, and misconception of knowledge (Wood, 2010). Instead, we work through and sometimes against the grain of maps as a way of creating alternative narratives, by and for the community and at a distance from those produced by the real estate sector. As Gutiérrez (2018) argues, the situated use of data that abound in counter-mapping projects provides opportunities for data activism.

As part of our counter-mapping approach, and in conjunction with housing justice community groups in Parc-Ex, PEAMP has developed maps to make visible the dramatic rise in evictions and to advocate for emerging struggles against rent hikes and displacement in Parc-Ex, as well as to support the rich community fabric in this neighbourhood. It was the care for the community that became particularly important in a context where the mistrust in institutions that are mandated to uphold the right to housing of vulnerable residents is at an all-time high. The project followed upon a longer-term collaboration and jointly co-authored a report with a network of community groups in Parc-Ex that was released in the first months of the COVID-19 pandemic to widespread media attention and community circulation (Parc-Ex Anti-Eviction Mapping Project, 2020).

Two online maps were developed and launched in 2021:[17] a map of evictions in Parc-Ex[18] and a Community Power Map.[19] Working closely with a housing justice and tenant support association in Parc-Ex, the *Comité d'action de Parc-Extension* (CAPE), PEAMP collaborated on co-constructing the first map based on data CAPE has been gathering in the course of their front-line support work for tenants. The joint aim is to document the dramatic rise in evictions in this area since 2017 (the first year that a rise in evictions became evident to CAPE), particularly in the political context of a government that denies that a housing crisis is in full swing (Bergeron, 2021). Given that wider eviction patterns are notoriously difficult to document (as often noted in literature on gentrification, for instance Chum (2015)), the aim with this map was to make visible the extent of evictions and to break the broader invisibility of the gentrification process due to the lack of public access to public data on evictions. In the absence of an open system to publicize evictions and the intricacies of the gentrification process, we do not claim the eviction map to be exhaustive (unlike many dominant approaches to mapping) but specify that it is merely the tip of the iceberg given that the map only shows the evictions of some residents who are willing to push back against their landlords. Nevertheless, the counter-mapping renders the effects on housing justice in the neighbourhood clear and publicly visible, while safe-guarding a degree of opacity to avoid overexposing vulnerable tenants to possible landlord reprisals (for instance, PEAMP did not publish exact points on the map, but approximate ballpark locations).

The second map, the Community Power Map, visualizes recent community actions and resistance, as well as the needs and desires expressed through their emergence. As community members and activists embedded in these issues in addition to being researchers, we were aware of the depth and value of community endeavours being conducted by residents, activists, and organizers in the neighbourhood. We did not want to focus solely on evictions in a decontextualized manner without also showing the agency, mutual aid, and active community power that undergirds the neighbourhood. While a range of community groups and residents were undertaking remarkable works to mobilize against gentrification, offer support, and increase public awareness of housing rights, these efforts were sometimes dispersed or invisible. Our Community Power Map aims to shed light on the importance of community work in Parc-Ex and create a resource to spatialize and characterize the various community efforts, struggles, victories, and capacities in the neighbourhood. More than being a representation of community work in the neighbourhood, the Community Power Map remained a tool of engagement and discussion. The map is used both as a resource depicting the stories of struggles and success in the neighbourhood, and a tool for igniting community exchange. The representational role of the map gave way to a more than representational one; the map became less important than the process of engaged activism that it ignited.

We have used maps to surface and make visible the landscapes of evictions and community work in Parc-Ex. However, as argued by Kitchin, Gleeson and Dodge (2013), maps do not always reflect the claims desired by their creators. The lives of maps continue to exist after their creation, and are being ceaselessly unfolded in different contexts shaped by 'personal, social, embodied, political and economic relations' (Kitchin et al, 2013: 494). The reliance on one visualization tool in one specific moment is not the goal of the project. Rather, the counter-mapping approach we have co-constructed is an ongoing, open process that situates data visualization and representation within a larger process as a vantage point and tool for discussion, mobilization, collaboration, and engaged activism. The maps are not instrumentalized as mere representation, but act as a point of reference and tool for igniting various talks, presentations, workshops, educational, and community exchanges. Alternative narratives of Parc-Ex are being re/created and revised, not only on our maps, but also in our public discourses and exchanges.

A key element and takeaway that anchors the counter-mapping approaches we have sought to put into practice is the importance of constructing thoughtful and sustainable community relays and infrastructures for their emergence. The creation and use of maps became possible due to an ongoing, durable community collaboration. It continues to live and be accountable to the social context for which these maps and discussions were being created. In emphasizing the importance of building durable community collaborations and data co-creation, we also foreground the less visible but no less important work of building social infrastructures and community relays to undergird effective and accountable data visualization and data activism.

Data activism through community research: CBAR, Digital Divides and Parc-Ex Curry Collective – Alex Megelas and Leonora King

Community relays, research accountability, and data activism can take multiple forms of alliance between researchers and community organizers. While, as seen earlier, the intersection of these practices centres data as a resource and a tool for counter-power, the solutions developed for the problems tackled are not necessarily technological in nature. In this section, we consider the practice of research as institutional confrontation and the extent to which the Parc-Ex community-based action research network (CBAR) proved to be an effective means of challenging institutional narratives relating to social impact stemming from the AI environment in Montreal, and advocated, via the Digital Divides research project (2022) for increased specificity in naming and documenting social impact. At the same time, importantly, the project created the conditions for autonomous, grassroots solutions.

The Parc-Ex community-based action research network (CBAR), coordinates the presence of university stakeholders in the neighbourhood, according to relational commitments that commit to reciprocity across all aspects of a research project. The network is autonomous and is not linked to any one university. It holds meetings every two months, at which area residents, local organizers, and students and researchers can attend to consider potential collaborations.

In 2019, at the request of local residents, an attempt was made by the network to reach out to noted Université de Montréal AI researcher Yoshua Bengio, who had led creation of the Montreal Declaration for a Responsible Development of Artificial Intelligence[20] to see if he might be willing to influence the Université de Montréal and the Campus MIL project towards the development of a housing mitigation strategy for the impact of its students on the neighbourhood. After meetings with Bengio's private firm, Element AI, came to a standstill, the CBAR network opted to expand on some of these exchanges as part of Digital Divides, a research project that committed to considering the impact on housing by the AI entrepreneurial sphere.

The report explored solutions to the inequitable distribution of the economic benefits of Montreal's AI sector and the evident economic disconnect between Montreal's AI tech hub, and the adjacent Parc-Ex neighbourhood. The report included recommendations on alternative housing models, collective benefit agreements, and AI Commons, in addition to smaller-scale, grassroots efforts. These initiatives aim to empower communities affected by the development of the AI sector while also democratizing its top-down governance structure. The researchers behind Digital Divides conclude the report by calling for a number of reforms to AI and housing policy that prioritize justice-oriented, rather than for-profit aims and involve meaningful community consultation. The report highlights the importance of social movement building – rather than just academic and government intervention – as key forces in the pursuit of these goals.

In an effort to address some of the systemic inequalities facing Parc-Ex residents, one of our community-based researchers, Leonora King, co-created a neighbourhood-led mutual aid initiative that sought to move resources from more privileged Montreal residents to vulnerable ones. Parc-Ex is home to many newcomers to the country who face economic, employment, and language barriers. Through her connections to community organizations in Parc-Ex and her close ties to residents themselves, Leonora founded the Parc-Ex Curry Collective (PECC). PECC provides a means of financial support to asylum-seeking women living in Parc-Ex by hiring them as chefs for catering and food distribution services. The Collective essentially empowers women who face economic barriers by leveraging one of their best assets – their cooking skills. PECC is based on a model of redistribution of resources in two ways: (1) more affluent residents can purchase meals made by asylum-seeking women, providing them with a source of regular income, and (2) some of the meals prepared go back into the community to feed more vulnerable residents who struggle with food insecurity. The Curry Collective partners with the Welcome Collective to identify vulnerable newcomer families who could use some warm, home-cooked meals. In terms of real impact, women facing employment barriers gain some financial autonomy, more vulnerable residents receive free meals and Montreal residents can sample authentic South Asian cuisine while contributing to a mutual aid network. The initiative continues to be an evolving grassroots solution to systemic barriers, enabling the integration of newcomers, raising awareness about the realities of asylum seekers and the barriers they face, and promoting equity by shifting resource allocation. In terms of longer-term impact, PECC chefs can rely on a stable source of income, enabling them to be more engaged residents as well as active participants of positive change.

The relationship between the Digital Divides project and the PECC demonstrated a new form of relationship within the sphere of university community partnership creation in the Parc-Ex community, and allowed for a deeper range of calls for socio-economic equity stemming from the scholarship which it furthered. The alignment between a research project and a community-led community economic development initiative, allowed for the creation of consistent brokering sites, through which relationality could be explored between neighbourhood residents and organizers, and researchers. It also led university students and researchers to commit to a broader range of outputs to academic work – through grounded participatory field work, through the creation of advocacy tools alongside an academic paper. In so doing, it nuanced and enriched the definitions of a kind of scholarship that can more fully and organically exist in the realms of organizing and advocacy.

DISCUSSANT RESPONSE

The nature of spatialized data power and politics – Rob Kitchin[21]

In recent years, data power and data politics have become key frames of analysis for making sense of the ways in which data and their associated systems and infrastructures are produced within, at the same time as they (re)produce, social and spatial relations. Power is vested in data through what they denote and how they can be used to reveal aspects of the world and inform decision-making (Beer, 2019). The lifecycle of data unfolds within socio-technical relations that can be highly contested, with political struggles over the means of production (ownership and control of systems, platforms, infrastructures), what is captured, in what forms, and how the resultant data circulate and are used (Kukutai and Taylor, 2016; Cheney-Lippold, 2017; Ruppert et al, 2017). These struggles are informed by ideological views and political narratives, and the work of data, their infrastructures, and data actors takes place within and helps shape political economies (Sadowski, 2019; Zuboff, 2019). In other words, increasing attention is being paid to the contested ways in which data and their attendant socio-technical assemblages of actors and actants are (re)produced, whose interests they serve, how data power in its manifest forms is exerted in practice, and how expressions of power are contested in practical, political, and normative ways through data activism and claims for data sovereignty, citizenship, and justice (Hintz et al, 2019; Dencik et al, 2022; Kitchin, 2022).

The short interventions in this section add to the ongoing work under way in digital geographies concerning the spatial politics and geographies of data power (for example, Ash et al, 2018; Graham and Dittus, 2022; Thatcher and Dalton, 2022). The essays provide a scalar and geopolitical analysis of the clustering of AI industries in Parc-Ex, Montreal, from their embedding in the global Internet, their siting in a global digital economy in which geopolitical rivalries between the US, China, and the EU are playing out, their reliance on extractive resources, and their framing with respect to environmental and climate concerns their grounding within a city-region and the effects on a neighbourhood community, and the embodied data subjectivities of residents and the limits to their data sovereignty. Collectively, they make clear the scaled, material, and contested politics of digital infrastructures, in which there are alliances and contestations between state, businesses, and communities, and the technologies themselves become the means of resisting data power and conducting politics through counter-data actions. Moreover, they highlight that data power is thoroughly interwoven into other political and power struggles: as with smart city developments, the AI ecosystem in Montreal is imbricated with real estate, property development, and rentier capitalism, and is a driver of urban-regional restructuring, housing crises, and spatial divisions of labour (Shelton et al, 2015; Kitchin et al, 2019). Data power and politics cannot be understood in isolation, but as contingent, relational, and contested relations that are intwined with other markets and vested interests constituting a complex 'power geometry' (Massey, 1993).

The analysis does useful work then in detailing the spatialized nature of data politics and power through a largely descriptive account of what is occurring and how expressions of data power are being resisted (in part, a function of consisting of seven short pieces). To take this analysis further, the varied and contested nature of concepts such as 'data power', 'technical territories', 'algorithmic violence', 'data sovereignty', and 'data activism' that are evoked could be examined, as could how they help make sense of the socio-spatial processes in action. Data power is noted to consist of hegemonic power and data counter-power (the authors' third category of 'powering data' refers to how data power is dependent, in a secondary sense, on resource extraction, rather than to the nature of power per se). Yet, power is a highly contested concept theoretically, its nature and operation diversely understood across philosophical viewpoints (Avelino, 2021). Such diverse conceptualization extends to understanding how power is spatialized (Allen, 2003; Coleman and Agnew, 2018). Moreover, at a practical level, power is expressed in varying ways (for example, domination, coercion, co-option, self-discipline, punishment, modulation, intervention, mediation, coordination, direction) and countered through diverse resistance tactics (for example, protest, non-compliance, disobedience, and so on). And power is mobilized and expressed in context; in this case, within a capitalist, neoliberalized, racialized, and gendered political economy and entrepreneurial urbanism. It is not enough, then, to say data power is at work without unpacking in nuanced theoretical terms how power is understood to work, through what means, and to what ends.

Similarly, the response to data power might be further elaborated by moving beyond political action and the noting of data activism to its normative framing. What should be the normative grounding of data counter-power? The pieces allude to data sovereignty and a sense to which citizens should have some level of autonomy in how data shape their lives. But what notion of sovereignty is being evoked? Other useful concepts might be data citizenship (rights, entitlements, and obligations delimited through data and with respect to data; Isin and Ruppert, 2015) and data justice (that people are treated in just, fair, and equitable ways by data-driven processes and systems; Dencik et al, 2022). Like power, each of these concepts is diversely understood and it is not enough to say we want data justice. For example, egalitarian data justice seeks equal treatment for all citizens, utilitarian data justice prioritizes the greatest good for the greatest number, and libertarian data justice contends that the free market is inherently just and people are treated as they deserve or can afford (Kitchin, 2022). What kind of data justice does and should underpin the data counter-power enacted in Montreal and the other interconnected sites that enable the city's AI cluster?

The pieces then provide a useful overarching analysis of the spatial and scalar politics of data power and my response is a call to extend and deepen this kind of analysis in order to provide a more theoretically nuanced explanation of the socio-spatial processes and emergent power geometries at play.

Acknowledgement
The analysis in this commentary was funded by European Research Council Grant no. 101052998, titled 'Data Stories'.

Notes
1   Université de Montréal
2   Concordia University
3   Université du Québec à Montréal
4   Université McGill, Afrique au Féminin
5   Concordia University
6   Concordia University
7   European University Institute
8   Concordia University
9   Minderoo Centre for Technology and Democracy, University of Cambridge

[10] Université de Montréal

[11] Discussant

[12] Facilitator

[13] For a history and critique of the ideology and impact of the ecosystem metaphor see: Krivý (2023), Norris and Suomela (nd), as well as the Digital Divides Project (2022).

[14] For a comprehensive analysis of the policies, financing mechanisms, and actors within the Canadian AI industry, see Brandusescu (2021).

[15] Similar processes have been observed in other 'tech hubs', like San Francisco and the Bay Area (Opillard, 2015; Stehlin, 2016; Maharawal and McElroy, 2018).

[16] https://antievictionmontreal.org/en/maps/

[17] https://antievictionmontreal.org/en/maps/

[18] Eviction Map of Parc-Extension – Interactive Version. Available at: https://antievictionmontreal.org/en/maps/evictions-map/

[19] Community Power Map of Parc-Extension – Interactive Version. Available at: https://antievictionmontreal.org/en/maps/community-power-map/

[20] https://recherche.umontreal.ca/english/strategic-initiatives/montreal-declaration-for-a-responsible-ai/

[21] Maynooth University Social Sciences Institute, Ireland

References

Allen, J. (2003) *Lost Geographies of Power*. Oxford: Blackwell.

Amodei, D. and Hernandez, D. (2018) 'AI and compute'. Open AI, 16 May. Available at: https://openai.com/blog/ai-and-compute/#fn2

Ash, J., Kitchin, R., and Leszczynski, A. (eds) (2018) *Digital Geographies*. London: Sage.

Avelino, F. (2021) Theories of power and social change: power contestations and their implications for research on social change and innovation. *Journal of Political Power*, 14(3): 425–448.

Baur, A. (2023) European dreams of the cloud: imagining innovation and political control. *Geopolitics*. https://doi.org/10.1080/14650045.2022.2151902

Beer, D. (2019) *The Data Gaze: Capitalism, Power and Perception*. London: Sage.

Beraldo, Davide and Milan, Stefania (2019) From data politics to the contentious politics of data. *Big Data & Society*, 6(2): 2053951719885967. https://doi.org/10.1177/2053951719885967

Bergeron, Patrice (2021) 'Parler de "crise a des impacts", selon François Legault'. *La Presse*, 6 May.

Brandusescu, A. (2021) 'Artificial intelligence policy and funding in Canada: public investments, private interests'. Centre for Interdisciplinary Research on Montreal, McGill University. Available at: www.mcgill.ca/centre-montreal/files/centre-montreal/aipolicyandfunding_report_updated_mar5.pdf

Bratton, Benjamin H. (2015) *The Stack: On Software and Sovereignty*. Cambridge, MA: MIT Press.

Brodie, P. (2020) 'Stuck in mud in the fields of athenry: Apple, territory, and popular politics'. Culture Machine. Available at: https://culturemachine.net/wp-content/uploads/2021/04/8.-Patrick-Brodie-revised-1.pdf

Brodie, P. (2023) Data infrastructure studies on an unequal planet. *Big Data & Society*, 10(1). https://doi.org/10.1177/20539517231182402

Broeders, D., Cristiano, F., and Kaminska, M. (2023) In search of digital sovereignty and strategic autonomy: normative power Europe to the test of its geopolitical ambitions. *Journal of Common Market Studies*, 61(5): 1261–1280. https://doi.org/10.1111/jcms.13462

Bruhn, J. (2014) Identifying useful approaches to the governance of indigenous data. *International Indigenous Policy Journal*, 5(2). https://doi.org/10.18584/iipj.2014.5.2.5

Bueger, Christian and Liebetrau, Tobias (2021) Protecting hidden infrastructure: the security politics of the global submarine data cable network. *Contemporary Security Policy*, 42(3): 391–413.

Campling, Liam and Colás, Alejandro (2021) *Capitalism and the Sea: The Maritime Factor in the Making of the Modern World*. London and New York: Verso Books.

Carroll, S. R., Garba, I., Figueroa-Rodríguez, O. L., Holbrook, J., Lovett, R., Materechera, S., et al (2020) The CARE Principles for Indigenous Data Governance. *Data Science Journal*, 19: article 43. https://doi.org/10.5334/dsj-2020-043

Chen, J. (2016) 'A strategy for limits-aware computing'. Computing Within Limits. Available at: https://computingwithinlimits.org/2016/papers/a1-chen.pdf

Cheney-Lippold, J. (2017) *We Are Data: Algorithms and the Making of Our Digital Selves*. New York: New York University Press.

Chum, A. (2015) The impact of gentrification on residential evictions. *Urban Geography*, 36(7): 1083–1098. https://doi.org/10.1080/02723638.2015.1049480

Coleman, M. and Agnew, J. (eds) (2018) *The Handbook on the Geographies of Power*. Chelmsford: Edward Elgar.

Conrad, Margaret (2022) *A Concise History of Canada*. Cambridge: Cambridge University Press.

Couture, S. and Toupin, S. (2019) What does the notion of 'sovereignty' mean when referring to the digital? *New Media & Society*, 21(10): 2305–2322. https://doi.org/10.1177/1461444819865984

Crawford, K. (2021) *Atlas of AI*. New Haven, CT: Yale University Press.

Dalton, C. M. and Stallmann, T. (2018) Counter-mapping data science. *The Canadian Geographer/Le Géographe Canadien*, 62(1): 93–101.

de Onís, Catalina M. (2021) *Energy Islands: Metaphors of Power Extractivism, and Justice in Puerto Rico*. Oakland, CA: University of California Press.

de Vries, A. (2023) The growing energy footprint of artificial intelligence. *Joule*, 7(10): 2191–2194.

Dencik, L., Hintz, A., Redden, J., and Trere, E. (2022) *Data Justice*. London: SAGE.

Desbiens, C. (2013) *Power from the North: Territory, Identity, and the Culture of Hydroelectricity in Quebec*. Vancouver: University of British Columbia Press.

Digital Divides Project (2022) 'Digital Divides – The Impact of Montreal's AI Ecosystems on Parc Extension: Housing, Environment and Access to Services'. CBAR Network in Parc-Extension, April.

Donaghy, T., Henderson, C., and Jardim, E. (2020) 'Oil in the cloud: how tech companies are helping big oil profit from climate destruction'. Greenpeace USA. Available at: www.greenpeace.org/usa/reports/oil-in-the-cloud/

Duarte, M. E. (2017) *Network Sovereignty: Building the Internet across Indian Country*. Seattle, WA: University of Washington Press.

Exentus, R. (2022) Le Comité d'action de Parc-Extension: En lutte pour le droit au logement. *Nouveaux Cahiers du Socialisme*, 27: 119–127.

Fox-Martin, A. (2022, October 11) 'Advancing digital sovereignty on Europe's terms'. Google Cloud, 11 October. https://cloud.google.com/blog/products/identity-security/advancing-digital-sovereignty-on-europes-terms

Foxworth, R. and Ellenwood, C. (2023) Indigenous peoples and third sector research: indigenous data sovereignty as a framework to improve research practices. *VOLUNTAS: International Journal of Voluntary and Nonprofit Organizations*, 34(1): 100–107. https://doi.org/10.1007/s11266-022-00458-7

Graham, M. and Dittus, M. (2022) *Geographies of Digital Exclusion: Data and Inequality*. London: Pluto.

Guay, E. and Drago, A. G. (2023) Liens sociaux, initiatives collectives et processus conflictuels: le cas des mobilisations contre la gentrification à Parc-Extension, Montréal. *Recherches Sociographiques*, 64(1): 119–141.

Guay, E. and Megelas, A. (2021) 'Le droit à la ville face à la gentrification des quartiers populaires montréalais. Une analyse des mobilisations à Parc-Extension et à Pointe-Saint-Charles'. In: J. Durand Folco (ed) *Montréal en chantier: Les défis d'une métropole pour le XXIe siècle*. Montreal: Éditions Écosociété, pp 216–229.

Gutiérrez, M. (2018) *Data Activism and Social Change*. London: Palgrave Macmillan.

Haché, Alex (2017) *Technological Sovereignty*, Vol 2. Barcelona. Available at: https://sobtec.gitbooks.io/sobtec2/en/

Haggart, B., Schoolte, J. A., and Tussikov, N. (2021) 'Introduction: Return of the State?' In: B. Haggart, N. Tuskov, and J. A. Scholte (eds) *Power and Authority in Internet Governance Return of the State?* London: Routledge, pp 1–12.

Hao, K. (2020) 'Tiny AI'. *MIT Technology Review*, 2 April. Available at: www.technologyreview.com/technology/tiny-ai/

Harvey, David (2001) Globalization and the 'Spatial Fix'. *Geographische Revue*, 2: 23–30.

Hilty, L. M. (2015) 'Computing efficiency, sufficiency, and self-sufficiency: a model for sustainability?' Computing Within Limits. Available at: https://computingwithinlimits.org/2015/papers/limits2015-hilty.pdf

Hintz, A., Dencik, L., and Wahl-Jorgensen, K. (2019) *Digital Citizenship in a Datafied Society*. Cambridge: Polity Press.

Hogan, Mél (2015) Data flows and water woes: the Utah Data Center. *Big Data & Society*, 2(2). https://doi.org/10.1177/2053951715592429

Isin, E. F. and Ruppert, E. (2015) *Being Digital Citizens*. London: Rowman & Littlefield.

Johnson, Alix (2023) *Where Cloud Is Ground: Placing Data and Making Place in Iceland*. Oakland, CA: University of California Press.

Jolivet, V., Reiser, C., Baumann, Y., and Gonzales, R. (2022) Before displacement: studentification, campus-led gentrification and rental market transformation in a multiethnic neighborhood (Parc-Extension, Montréal). *Urban Geography*, 44(5): 983–1002.

Kee, Robert (2000) *The Green Flag: A History of Irish Nationalism*. Penguin UK.

Keller, J. B. (2023) 'The next superpower battlefield could be under the sea in Africa'. *Foreign Policy*, 10 March. Available at: https://foreignpolicy.com/2023/03/10/africa-china-russia-subsea-cables/

Kitchin, R. (2022) *The Data Revolution: A Critical Approach to Big Data, Open Data, and Data Infrastructures*. 2nd edition. London: Sage.

Kitchin, R., Cardullo, P., and di Feliciantonio, C. (2019) 'Citizenship, Social Justice and the Right to the Smart City'. In: P. Cardullo, C. di Feliciantonio, and R. Kitchin (eds) *The Right to the Smart City*. Bingley: Emerald, pp 1–24.

Kitchin, R., Gleeson, J., and Dodge, M. (2013) Unfolding mapping practices: a new epistemology for cartography. *Transactions of the Institute of British Geographers*, 38(3): 480–496.

Krivý, M. (2023) Digital ecosystem: the journey of a metaphor. *Digital Geography and Society*, 5. https://doi.org/10.1016/j.diggeo.2023.100057

Kukutai, T. and Taylor, J. (eds) (2016) *Indigenous Data Sovereignty: Toward an Agenda (CAEPR)*. Canberra: Australian National University Press.

Kwet, Michael (2019) Digital colonialism: US empire and the new imperialism in the Global South. *Race & Class*, 60(4): 3–26.

Lehuedé, Sebastián (2022) Territories of data: ontological divergences in the growth of data infrastructure. *Tapuya: Latin American Science, Technology and Society*, 5(1).

Lewis, J. E., Arista, N., Pechawis, A., and Kite, S. (2018) Making kin with the machines. *Journal of Design and Science*. https://doi.org/10.21428/bfafd97b

Li, P., Yang, J., Islam, M. A., and Ren, S. (2023) 'Making AI less "thirsty": uncovering and addressing the secret water footprint of AI models'. Available at: https://arxiv.orf/abs/2304.03271

Maharawal, M. M. and McElroy, E. (2018) The Anti-Eviction Mapping Project: counter mapping and oral history toward Bay Area housing justice. *Annals of the American Association of Geographers*, 108(2): 380–389.

Malmodin, Jens and Lundén, Dag (2018) 'The electricity consumption and operational carbon emissions of ICT network operators 2010–2015'. Stockholm, Sweden: Centre for Sustainable Communications.

Marks, Laura U., Clark, Joseph, Livingston, Jason, Oleksijczuk, Denise, and Hilderbrand, Lucas (2020) Streaming media's environmental impact. *Media+ Environment*, 2(1).

Massey, D. (1993) 'Power Geometry and a Progressive Sense of Place', in J. Bird, B. Curtis, T. Putnam, G. Robertson, and L. Tickner (eds) *Mapping the Futures*. London: Routledge, pp 59–69.

Mattoni, Alice and Odilla, Fernanda (2021) Digital media, activism, and social movements' outcomes in the policy arena. the case of two anti-corruption mobilizations in Brazil. *Partecipazione & Conflitto* 14: 1127–1150. https://doi.org/10.1285/i20356609v14i3p1127

Maxwell, Richard and Miller, Toby (2012) *Greening the Media*. Oxford: Oxford University Press.

McMahon, R., Chasle, A., and Whiteduck, T. (2015) First Mile methodologies in community informatics research: learning from First Nations. *The Journal of Community Informatics*, 11(3). https://doi.org/10.15353/joci.v11i3.2705

Moos, M., Revington, N., Wilkin, T., and Andrey, J. (2019) The knowledge economy city: gentrification, studentification and youthification, and their connections to universities. *Urban Studies*, 56(6): 1075–1092.

Mueller, Milton (2017) *Will the Internet Fragment? Sovereignty, Globalization and Cyberspace*. Cambridge: Polity Press.

Munn, L. (2023a) *Technical Territories: Data, Subjects, and Spaces in Infrastructural Asia*. Ann Arbor: University of Michigan Press.

Munn, L. (2023b) Red territory: forging infrastructural power. *Territory, Politics, Governance*, 11(1): 80–99. https://doi.org/10.1080/21622671.2020.1805353

Nobrega, C. and Varon, J. (2020) 'Big tech goes green(washing): Feminist lenses to unveil new tools in the master's houses'. Global Information Society Watch. Available at: www.giswatch.org/node/6254

Nora, S. and Minc, A. (1980) *The Computerization of Society. A Report to the President of France*. Cambridge, MA: MIT Press.

Nordenstreng, K. and Schiller, H. (1978) *National Sovereignty and International Communication: A Reader*. Norwood, NJ: Ablex Publishing.

Norris, T. B. and Suomela, T. (nd) 'Information *in* the ecosystem: against the "information ecosystem"'. *First Monday*. Available at: https://firstmonday.org/ojs/index.php/fm/article/view/6847/6530

Nungak, Z. (2017) *Wrestling with Colonialism on Steroids: Quebec Inuit Fight for their Homeland*. Montreal: Véhicule Press.

Onuoha, M. (2018, July 2). 'Notes on algorithmic violence'. Github. Available at: https://github.com/MimiOnuoha/On-Algorithmic-Violence

Opillard, F. (2015) Resisting the politics of displacement in the San Francisco Bay area: anti-gentrification activism in the tech boom 2.0. *European Journal of American Studies*, 10(3).

Ortiz Freuler, J. (2022) The weaponization of private corporate infrastructure: Internet fragmentation and coercive diplomacy in the 21st century. *Global Media and China*, 8(1): 6–23.

Pannier, A. (2021) 'The changing landscape of European cloud computing: Gaia-X, the French national strategy, and EU plans'. IFRI: Institut Français des Relations Internationales, France. Available at: https://policycommons.net/artifacts/1806934/the-changing-landscape-of-european-cloud-computing/2540689/ CID: 20.500.12592/zdbqmt.

Parc-Extension Anti-Eviction Mapping Project (2020) 'MIL façons de se faire évincer: The University of Montreal and Gentrification in Parc Extension'. Available at: https://antievictionmontreal.org/wp-content/uploads/2020/06/MIL-façons-de-se-faire-évincer-LUniversité-de-Montréal-et-la-gentrification-à-Parc-Extension.pdf

Pasek, Anne, Vaughan, Hunter, and Starosielski, Nicole (2023) The World Wide Web of carbon: towards a relational footprinting of ICT's climate impacts. *Big Data & Society*, 10(1). https://doi.org/10.1177/2053951723 1158994

Pasquinelli, M. and Joler, V. (2020) 'The Nooscope manifested: artificial intelligence as instrument of knowledge extractivism'. Available at: https://nooscope.ai

Pohle, J. and Thiel, T. (2020) Digital sovereignty. *Internet Policy Review*, 9(4). https://doi.org/10.14763/2020.4.1532

Powers, S. and Jablonski, M. (2015) *The Real Cyberwar: The Political Economy of Internet Freedom*. Chicago, IL: The University of Illinois Press.

Raghavan, B. (2015) 'Abstraction, indirection, and Sevareid's Law: towards benign computing'. Computing Within Limits. Available at: https://computingwithinlimits.org/2015/papers/limits2015-raghavan.pdf

Reese, E., DeVerteuil, G., and Thach, L. (2010) 'Weak-center' gentrification and the contradictions of containment: deconcentrating poverty in downtown Los Angeles. *International Journal of Urban and Regional Research*, 34(2): 310–327.

Renzi, A. and Irendra, Radjawali (forthcoming) The view from the blackbox: insurgent technopolitics in Indonesia's data and extraction economies. *Transversal Journal*, issue Technopolitics of Care.

Renzi, A., Vukov, S., Guay, E., Shahamati, S., Baumann, Y., Chen, S. and Emperador Badimon, M. (2022) Le Big Data contre le droit à un chez-soi? Le secteur de l'intelligence artificielle et l'embourgeoisement de Parc-Extension. *Droits et libertés*, 41(1): 31–33.

Rone, J. (2023) The shape of the cloud: contesting data centre construction in North Holland. *New Media and Society*, 0(0). https://doi.org/10.1177/14614448221145928

Ruiz, R. (2021) *Slow Disturbance: Infrastructural Mediation on the Settler Colonial Resource Frontier*. Durham, NC: Duke University Press.

Ruppert, E., Isin, E., and Bigo, D. (2017) Data politics. *Big Data & Society*, 5(2): 1–7.

Sadowski, J. (2019) When data is capital: datafication, accumulation, and extraction. *Big Data & Society*, 5(1): 1–12.

Saenko, K. (2020) 'It takes a lot of energy for machines to learn – here's why AI is so power-hungry'. The Conversation. Available at: http://theconversation.com/it-takes-a-lot-of-energy-for-machines-to-learn-hereswhy-ai-is-so-power-hungry-151825

Sanders, C. (2022) 'Microsoft cloud for sovereignty: the most flexible and comprehensive solution for digital sovereignty'. Microsoft Blog, 19 July. Available at: https://blogs.microsoft.com/blog/2022/07/19/microsoft-cloud-for-sovereignty-the-most-flexible-and-comprehensive-solution-for-digital-sovereignty/

Schwartz, R., Dodge, J., Smith, N. A., and Etzioni, O. (2019) 'Green AI'. ArXiv:1907.10597 [Cs, Stat]. Available at: http://arxiv.org/abs/1907.10597

Shelton, T., Zook, M., and Wiig, A. (2015) The 'actually existing smart city'. *Cambridge Journal of Regions, Economy and Society*, 8: 13–25.

Sprague, M. and Rantisi, N. (2019) Productive gentrification in the Mile-Ex neighbourhood of Montreal, Canada: exploring the role of the state in remaking urban industrial clusters. *Urban Research and Practice*, 12(4): 301–321.

Starosielski, N. (2015) *The Undersea Network*. Durham, NC: Duke University Press.

Startup Montréal (2021) 'Map for Montreal's artificial intelligence ecosystem'. Startup Montréal, 19 April. Available at: https://startupmontreal.com/en/our-map-for-montreals-artificial-intelligence-ecosystem/

Stehlin, J. (2016) The post-industrial 'shop floor': emerging forms of gentrification in San Francisco's innovation economy. *Antipode*, 48(2): 474–493.

Stockwell, Sarah E. (ed) (2008) *The British Empire: Themes and Perspectives*. Oxford: John Wiley & Sons.

Sustainability in the Digital Age (SDA) (2020) 'Digital disruptions for sustainability agenda (D^2S Agenda): research, innovation, action'. Future Earth. Available at: https://sustainabilitydigitalage.org/featured/wp-content/uploads/D%5E2S-Agenda-Report-2020.pdf

Sutherland, B. (2022) 'Strategies for degrowth computing'. Computing Within Limits. Available at: https://computingwithinlimits.org/2022/papers/limits22-final-Sutherland.pdf

Szeman, Imre (2019) *On Petrocultures: Globalization, Culture, and Energy*. Morgantown: West Virginia University Press.

ten Oever, Niels (2022) 5G and the notion of network ideology, or: The limitations of sociotechnical imaginaries. *Telecommunications Policy*, 47(5). https://doi.org/10.1016/j.telpol.2022.102442

Thatcher, J. and Dalton, C. (2022) *Data Power: Radical Geographies of Control and Resistance*. London: Pluto.

Toyama, K. (2015) 'Preliminary thoughts on a taxonomy of value for sustainable computing'. Computing Within Limits. Available at: https://computingwithinlimits.org/2015/papers/limits2015-toyama.pdf

Université de Montréal (2021) 'Sustainable Development at Université de Montréal, Strategic Plan'. Available at: www.umontreal.ca/public/www/images/developpement-durable/ANG_UdeM_Sunstainability_Strategy.pdf

Utrata, A. (2022) 'Amazon web services, cloud computing and corporate control'. *The Darwinian*. Available at: https://darwin-college-cambridge.shorthandstories.com/amazon-web-services-cloud-computing-and-corporate-control/index.html

Vaughan, Hunter, Pasek, Anne, Silcox, Nicholas R., and Starosielski, Nicole (2023) ICT environmentalism and the sustainability game. *Journal of Language and Politics*, 22(5): 640–660. https://doi.org/10.1075/jlp.22125.vau

Watts, Laura (2019) *Energy at the End of the World*. Cambridge, MA: MIT Press.

Winseck, D. (2019) 'Internet Infrastructure and the Persistent Myth of US Hegemony'. In: B. Haggart, K. Henne, and N. Tusikov (eds) *Information, Technology and Control in a Changing World: Understanding Power Structures in the 21st Century*. London: Palgrave Macmillan, pp 93–120.

Winseck, D. R. and Pike, R. M. (2007) *Communication and Empire: Media, Markets, and Globalization, 1860–1930*. (American Encounters/Global Interactions.) Durham, NC: Duke University Press.

Wood, D. (2010) *Rethinking the Power of Maps*. New York: Guilford Press.

Zewe, A. (2023) 'Computers that power self-driving cars could be a huge driver of global carbon emissions'. MIT News, 13 January. Available at: https://news.mit.edu/2023/autonomous-vehicles-carbon-emissions-0113

Zuboff, S. (2019) *The Age of Surveillance Capitalism: The Fight for the Future at the New Frontier of Power*. New York: Profile Books.

9

A Canon Is a Blunt Force Instrument: Data Science, Canons, and Generative Frictions

Aviva de Groot,[1] *George Fletcher,*[2] *Gijs van Maanen,*[3]
Akrati Saxena,[4] *Alexander Serebrenik,*[5] *and Linnet Taylor*[6]

Introduction

Spatially close, though worlds apart. The contributors to this commentary – 'we'; 'us' – conduct research and teach on data and technology-related issues at three Dutch universities. Some of us work at the same departments and teach in the same programmes. We bump into one another during our daily commutes and replenish our energy levels with the help of the same coffee machines after our lectures. We talk, and sometimes even discuss our research with one another. But do we also understand each other? What would that even mean? When we talk about 'data', do we talk about the same thing? Is that even necessary? What does 'science' for each of us entail? What does this mean for the education we collectively provide? What is the direction – scientifically, ethically, politically, professionally, commercially – the Bachelor programmes we are all involved in head toward?

National science policy in the Netherlands, as well as at the level of universities themselves, tends to prioritize in various ways computer and computational sciences over the social sciences and humanities (Taylor et al, 2023). We feel that the oppositions that are produced and reinforced through such policies are both false and unproductive, and this collective uneasiness motivated some of us to initiate a conversation about what it would mean to think and work together. How do our academic lives 'hang together' (Mol, 2014) beyond our encounters near coffee machines in the hallways, and our names on the timetables the students would find when logging in to their university pages?

When asking these and many other questions, we realized that we lacked the language, a common vocabulary, not only to answer the questions with which we started, but also to ask them.

Not only did many of the key concepts used in our research and education – data, algorithm, ethics, ontology, law – mean and do different things for all of us, but concepts indispensable to some – for example, justice – would be non-existent in the disciplinary universe of others.

We therefore needed to take a step back and reflect on how to have a conversation without sharing a common language. Our provisional solution was to take what we dubbed as 'canonical objects' as the focal points in our discussions. We borrow the notion of the canon from literary criticism, where it is used to mean a body of literature that over time comes to be taught as defining a particular culture (Bloom, 1994). For this reason, the canon has also been the focus of decolonial critics, who argue that we should critically interrogate the hegemonic discourses of Western culture (Spivak, 1990).

Based on this notion, we started to analyse concepts which each of us consider conceptually stable enough in our different disciplines that they might be taught on a Bachelor's-level course. In other words, we took our disciplinary backgrounds and educational responsibilities as conversational starting points. Our roughly defined meta-question was how our disciplinary backgrounds produced different conceptions of the same terms, how these differences could be generative or problematic, and how our disciplines become invested in a particular interpretation?

What we called canonical objects is also strongly related to how some of us used and understood the notion of boundary objects. A classic definition of boundary objects is that these 'have different meanings in different social worlds but their structure is common enough to more than one world to make them recognizable, a means of translation' (Star and Griesemer, 1989). Boundary objects thus allow different 'social worlds' to work together without requiring them to be able to (completely) understand one another. *If* our canonical objects would indeed function like boundary objects, we would have to find out and explicate in what way we would be working together, and how these concepts help us do that.

As part of our exploration, we also include answers from the generative large language model (LLM) ChatGPT3.5. This LLM draws on Internet content, and therefore offers a generalized and social version of the canon, replicating the most common tropes about our chosen objects of study available online. We wanted to include these tropes as a demonstration of how interdisciplinarity will often, if not theoretically informed, mash up disciplinary perspectives in ways that produce artificial neutrality and consensus on issues of real incompatibility, friction, and tension. As such, we aim to sensitize the reader to the incentives present in our different academic environments to focus on the lowest common denominator in debates on socio-technical phenomena, and to minimize disagreement.

Furthermore, interesting both conceptually and practically, was and still is, our attempt to create some level of mutual understanding (Gadamer, 2014), potentially with the help of boundary objects whose functioning depends on a *lack* of mutual understanding. How does our attempt to foster understanding about how we hang together or not, change our collaborations? What does this attempt do to the canonical objects that we used as conversational lubricants? How, to put that differently, does discussion and explicating our disciplinary divisions, change our capacities to, for example, teach together? And subsequently, what are generative but also less and non-generative ways of *dis*agreeing with one another?

In this contribution we present the results of the conversation we have had so far about two canonical concepts: artificial intelligence ('AI') and 'trust'. Together we made a list of *potential* canonical concepts (see the Appendix) – so concepts that would be taught in a BSc/BA programme/ course – and from this list picked two of those with the most multifaceted disciplinary usage to discuss here. Each of us was asked to briefly explain how from their (disciplinary) point of view the concept was understood and taught in our undergraduate programmes. These brief reflections are accompanied by statements about our own positionality (Harding, 1991; Haraway, 1991) in which each of us situates her/himself in the academic tradition in which they were educated. We have included these because we presumed that academic disciplines (and what have been termed signature pedagogies (Poole, 2009)) were and still are the key factors that influence the types of academic social worlds most of us live in. In the discussion we present some of the themes that emerged in our conversation, and that help to understand how our academic activities hang together – or not.

Studying and teaching AI

AI, data systems, and George[7]

> *You wouldn't believe how difficult it is to get rid of philosophy. Not to philosophise is still to philosophise, but without method, with a sort of naive brutality.*
>
> Gabriel Séailles 1904

'Intelligence' has historically played a critical role in the evolution of the study of data systems in the computing science (CS) and data science (DS) disciplines. It can even be argued that the rise of the concept itself tracks the birth of these disciplines; it is one of the core ideas which enabled the emergence of these disciplines in the 20th century. I briefly highlight two canonical roles here; canonical in the sense that they have informed what is traditionally accepted as part of the common culture of these disciplines and, in particular, that of the data systems research community.

The first is the orientation towards and interest in intelligence itself and in the particular ways in which the concept has been understood. Our conceiving of Intelligence, and preoccupation with realizing this understanding, is our intellectual inheritance of Leibniz (for example, his idea of mechanical rationality, as famously articulated as 'Calculemus!'), Boole (for example, the logical and probabilistic framings of his Universal Laws of Thought), and their many fellow workers in the laying of the foundations and scoping of CS and DS prior to the 20th century (C. Babbage, C. S. Peirce, R. Llull, and others). The logical, rational, mathematical, and mechanical framings of what counts as Intelligence articulated by Leibniz, Boole, and company (namely, that there is something called Intelligence, that it is a feature of brains, and that brains are computers) set the stage (and the boundaries) for the CS/DS discipline's understanding of what meaningful questions we might ask about the world. These framings permeate the vocabularies and worldviews we inhabit in our research literature and in the classroom. Boole's division between logical and probabilistic reasoning set the stage for the two traditional camps of the discipline of AI, the so-called 'neats', developing symbolic reasoning, and the 'scruffies', developing inductive (probabilistic, statistical) reasoning. This is a structuring which still informs much of the work of AI and the related study of machine learning (ML).

The second is how this conceiving of and preoccupation with Intelligence is normalized (and thereby forgotten) in the broader field of computing and data science. The history of AI is very much the history of computing and data science, and within this the study of data systems; techniques and viewpoints developed first in AI eventually make their way into the rest of CS and DS as techniques and viewpoints in the day-to-day toolkits of researchers in areas beyond AI. My own area of study is data systems, the systems for the care, protection, and effective use of collections of data. A central aspect of any data system is the languages for interacting with the data collection, for articulating information needs. At the heart of these languages are capabilities for expressing the logical structure of data (for example, finding connections in a social network between people and the places they work) and the statistical structure of data (for example, determining the average salary at each such workplace). These capabilities are so mundane that we no longer note (or rather, are essentially completely unaware of) the intellectual roots of this orientation to datafy the world and then to understand it through these capabilities. Yet the analyses enabled by these capabilities, these 'universal' capabilities of 'thought' articulated by Boole and company, are at the base of the current developments in AI and ML.

Of course, canons evolve as fields evolve. There is building momentum within several areas of CS/DS towards a 'humanistic turn', placing people in the centre of our work, moving beyond our Leibnizian and Boolean notion of Intelligence and the harmful aspects of the shadow it casts over the study of data systems.[8] There are many signals of this turn. As an example, a recent experience I had in conversation with graduate students: students in the field are discovering and recommending to their advisers the work of early pioneers in the study of computing as a socio-technical phenomenon, such as Philip Agre (1995, 1997). Perhaps AI and data systems will gradually reconfigure their relationships in order to make more space in the canon for actual people (in place of the 19th-century Intelligence of AI), for creating space to study what it is we are doing together with data technologies and what we do to each other with these technologies and how we might do things differently and better, for people-centred notions of better.

AI, law, and Aviva[9]

Human intelligence is typically understood in law as the capacity to act in a 'rationally' informed way. It is of canonical legal concern in a very broad range of social situations, wherein what counts as rational depends on the specific legal interest in question. Law, for example, engages in what it sees as physiological approaches of cognitive agency for the assessment of human actors' accountability, culpability, or responsibility. For example, age, health (including mental) state, dis/ability are factored in rules around financial and other transactions, tort and criminal culpability, and the right to make medical decisions about oneself. The law's inherent responsiveness to social and political developments means it also gets lobbied to engage with, for example, novel neurological insights. Examples exist in the assumed moral and temporal 'shortsightedness' of the adolescent brain, and other neurological approaches of free will and autonomy that are debated in the field of 'neurolaw' (for example, Bigenwald and Chambon, 2019). Problematic historical examples of such responsiveness lead to the dark waters of craniometry, and to law's dealings with flawed intelligence-related findings from the field of genetics (Beckwith and Pierce, 2018).

Less physiological approaches of intelligence, for example, exist in the law's framing of the capabilities of 'average persons' with regard to their legal, medical, bureaucratic, and lately, technological literacy, translating into legal notions on how much, and what kinds, of information individuals 'reasonably' need in order to participate responsibly in medical, administrative, and other decision-making. Law's focus here is on the ability of persons to act in their self-interest. Intelligence, in such framings, tends to attach to privilege and to notions of merit and desert. First, the knowledge, experiences, and information needs of less privileged groups typically do not inform the 'average person' benchmarks (de Groot, 2023). Second, persons in need of (state) support are not 'lawyered' into information positions that help them to act in their self-interest – quite the opposite (for example, Bridges, 2017; Eubanks, 2018).

In short, the law is interested in framing intelligent human agency in order to govern situations in which it matters morally to do so. It works with theories, assumptions, and benchmarks for this. As signalled in preceding paragraphs, the 'North-Western' traditions that I teach in have inherited problematic colonialist and other ideological thought with regard to their modelling, and mainlining, of the individual, rational, reasoning, intelligent subject (Wright, 2001; D'Souza, 2018; Moyn, 2019; Katz, 2020). Corresponding rights (to vote, to perform political functions, enter into financial relations, decide about one's own body, not be discriminated against in education and career, not to be 'wrongly manipulated' in commercial advertising) have suffered as a result.

These are important themes when looking at how the law approaches what is called '*artificial* intelligence' and how it chooses theories, assumptions, and benchmarks to work with. For example, law tends to let itself be informed by the tech industry's own framings around complexity and explainability to inform, for example, transparency and explanation obligations that should help the intelligent, autonomous subject act in their self-interest when they are affected by AI. I like to alert students to how historically, framings around *medical* complexity were used by physicians many decades ago to argue against informed consent regulations, and as I will argue under the 'trust' lemma the same arguments are used today to mystify 'AI' (Katz, 1984; de Groot, 2023).

More generally, historically problematic ideological alignments between law and the AI fields have been researched from various disciplinary angles (for example, Benjamin, 2019; Katz, 2020; Lepore, 2020; Broussard, 2023), and some in technology law are acknowledging the need to engage with these insights.[10] Teaching (from) this critical angle is not standard practice in law schools in the Netherlands however, which are generally known to be conservative. This stands to impoverish legal education on digital technologies while courses on this are slowly being added to Bachelor curricula. The mentioned multidisciplinary histories of AI can usefully inform our 'budding canonical' teaching on 'AI' moving forward, and hopefully incentivize future scholars towards useful interdisciplinary engagements.

AI, political philosophy, and Gijs[11]

Just as doing philosophy is not a coherent and easily identifiable and definable activity, answering the question of how philosophers in general, or *political* philosophers, define AI, is a task designed to fail. I will nevertheless try to present some generalizations that hopefully capture the work that some philosophers, especially those working on politics, sometimes do. First, it is good to note that 'AI' and 'algorithms' are not always neatly distinguished from one another by philosophers. AI and algorithms have something to do with computers and machines that are engaged in activities that potentially relocate some capacities for action normally present in humans, to the realms of commerce and technology.[12] Whether or not this is problematic depends on how one thinks who or what is now in charge of our lives: the smart algorithms themselves, or the businesses who own them. Regardless of the specific direction an answer to this question takes, AI and algorithms pose or *could* pose fundamental problems to our human condition according to many philosophers (philosophers love problems).[13] Second, and third, what AI and algorithms are or do – apart from putting pressure on our human and societal capacity for action – is sometimes considered to be less important than the moral-political fact that they (could) present societal harms.[14] To put that differently: (political) philosophers often approach AI and algorithmically related problems through the moral and political lenses that they have been grinding the last 2,000 years, rather than starting their study of AI and algorithms in practice, and empirically. The conceptual and methodological toolkits philosophers use, offer them the ability to present fundamental moral-political critique of AI and algorithms (and sometimes also praise). They come, however, with the cost of presenting ambiguous and superficial analyses of the problems AI and algorithms pose in practice. One example to illustrate this point is the call to 'democratize' (the algorithms of) platforms such as Facebook because of the problematic character of Facebook's algorithm.[15] Arguments like these combine (a) philosophical theories on democracy, with (b) an idea that platforms use algorithms, and because of that, we should take back control, by (c) democratizing these algorithms. A possible limitation of such an approach is the conflation of extremely complex commercial ecosystems[16] and the presumed influence of 'algorithms' and 'AI' on what they do, potentially resulting in the wrong thing to take back control of (algorithms). It is sometimes a bit too easy and convenient to blame 'the mutant algorithm'.[17] Another difficulty concerns the indirect or implicit legitimization of platforms such as Facebook that goes hand in hand with the taking back control argument. To what extent is it of value to, indeed, take control over Facebook, and to what extent is democracy, or another normative theory, always applicable to supposedly problematic technology developments (see Dunn, 2005)?

AI, computer science curricula, and Alexander[18]

AI has a difficult relation with the traditional computer science curricula. For example, the model ACM Software Engineering curriculum (Ardis et al,

2015) does not mention it, while the 2020 Computing Curricula explicitly states that '[t]he study of artificial intelligence, an area of renewed interest, is not included in this report because an ACM/IEEE-CS sponsored curricular guideline does not currently exist'. Even curricula that do include courses on AI or related topics such as ML, tend to box them as a separate area of computer science endeavour not necessarily related to more traditional CS topics, such as algorithms, programming, or operating systems. AI courses within the traditional CS curriculum tend to avoid reflecting on the meaning of the terms 'artificial' or 'intelligence' in AI and use AI as an umbrella term covering several unrelated topics commonly associated with AI, such as knowledge representation, heuristic search algorithms, planning, ML, or neural networks. Is there a real difference between such computational tasks as sorting and planning, the first traditionally not being seen as an AI task, the second one seen as an AI task? Why is classification based on logistic regression not considered to be an AI task while classification carried out by a neural network is? In a way, traditional CS seems to treat AI as a metaphor: in the same way as Software Architecture describes main 'building blocks' of a software system, similarly to architecture describing actual building blocks, but is by no means subject to limitations of the physical world, AI is a study of tasks requiring 'intelligence', that is, tasks deemed to be too difficult for pre-AI approaches. However, metaphors have their limitations: revisiting the adage of Shimon Peres on pre-election polls, metaphors are 'like perfume – nice to smell, dangerous to swallow'. In other words, they have their charms, but one should be careful not to take them too literally. In a way, this triggers a fundamental question: does AI even exist as a coherent discipline?

Contemporary exposure of LLM-based techniques questions both the aforementioned vision as well as multiple elements of the traditional computer science curriculum. Do we still need to teach programming if LLMs can produce source code based on a series of natural language prompts? How would these models affect both the ways the software is going to look in the future and the ways it is going to be developed? The latter questions reflect the contemporary software engineering (SE) perspective on AI: similarly to philosophers, SE researchers are not interested in AI per se but rather in its impact on SE practices, trying either to understand how SE practices should be adjusted to address the challenges induced by the need to develop AI systems (SE4AI) or trying to understand how AI can be used or adjusted to support SE practices (AI4SE).

AI, computer science, and Akrati[19]

AI has a long and fascinating history in the field of CS. The origins of AI can be traced back to the 1950s, when computer scientists began exploring the idea of creating machines that could think like humans. Researchers such as John McCarthy, Marvin Minsky, and Claude Shannon laid the groundwork for AI with their groundbreaking work on topics such as logic and information theory. Around the 1970s, AI research focused on the development of expert systems, which were designed to mimic the knowledge and decision-making abilities of human experts in a particular field. Examples of these systems include MYCIN (Shortliffe, 2012), which diagnosed bacterial infections, and DENDRAL (Lindsay et al, 1993), which identified molecular structures. Later on, around the 1980s, the focus of AI research shifted towards ML, which involves training machines to learn from data.

When students study CS, they are offered a course on AI where they learn how to use programming, statistics, and robotics to design intelligent machines, also called expert systems or agents. CS students learn logic, programming, mathematics, data structure, algorithms, and software development in the starting years and then they learn AI through a combination of theoretical and practical coursework and also cover its applications in various fields. After acquiring theoretical knowledge, students start working on practical projects that involve developing AI systems. This involves using tools and libraries such as TensorFlow, PyTorch, and scikit-learn to create and train ML models. In these courses, students also learn how to handle large-scale datasets, how to process and clean data so that they can be used by AI systems. AI is a broad field, and students can specialize in various subfields of AI such as ML, Natural Language Processing (NLP), Robotics, and Computer Vision, and these areas are taught through specialized courses in Masters.

AI is a rapidly evolving field, and it is essential for students to keep up with the latest developments. They need to stay updated with new algorithms, tools, and libraries and continuously hone their skills through practice and experimentation. These days, universities are broadly promoting Bachelor and Master degrees in DS and AI, given the job market requirement. These courses focus on hands-on projects and practical exercise to give students a more practical understanding of the subject, as employers often look for candidates with practical experience and the ability to apply theoretical knowledge to real-world problems.

AI is not only used by the students who have a specialization or degree in it. These days, many people even from different backgrounds learn AI to apply it in their domain or learn it due to the big job market in this area. Most of these candidates focus on the practical aspect of AI. There are abundant online (most of them also free) resources and guidance that can help in gaining the knowledge and skills necessary to become successful AI practitioners. They mainly learn how to use AI APIs, data management tools such as SQL, Pandas, and if required then how to update the existing models based on the application requirements.

Back to our big question – 'What is AI?' In simple words, for computer and data scientists, AI is designing intelligent algorithms to automate predictive, repetitive, or decision-making tasks. As Alexander explained in the previous section, the concept of AI is closer to algorithms, as we design automation algorithms. AI has been used in many different areas from healthcare to finance to education and manufacturing. Most of the people designing AI systems are more involved in its practical application than its theoretical side.

AI, critical data studies, and Linnet[20]

Teaching data and AI governance to law students – one of the tasks I undertake as a teacher in critical data studies – means using a definition of AI that is usually very light on detail. The students are expected to work on legal problems rather understand the technology, so the understanding of what AI-based systems are actually doing is extremely thin. The main effect of this is that even when students have been provided with greater technical detail, the legal disciplines tend not to encourage them to integrate that technical understanding with their analysis of legal and regulatory problems. Often, however, in order to make sense of a problem sufficiently to understand what to do to remedy it, it is necessary to have an idea of what features of the technology are generating that problem and how.

For a researcher in critical data studies, AI exists in layers. At one layer, it is (as Crawford and Joler have described, 2018), a 'system made of human labor', where material goods and infrastructures are integrated into what theorists have termed an 'assemblage' (Deleuze and Guattari, 1988) – a collection of heterogeneous elements working together to produce something with specific social effects – which then performs computational tasks. In the next layer, that of data, it consists of training data sourced from the social world, which theorists have defined as bodies (Floridi, 2016; Sandvik, 2023), social relations (Mhlambi, 2020; Tierra Común, 2021), or labour (Fuchs, 2014), concretized into factors of digital production. At another layer, that of statistical computation, those factors of production are transformed into moving parts of an analytical model, which weights different pieces of information differently and uses them in conjunction and opposition to each other to produce outputs which are then produced through an interface in a way that can make sense to particular groups of humans. Finally, there is the business layer, where those outputs are transformed into either knowledge or financial capital, and the assemblage earns its living.

For critical data studies, if we cannot understand who is involved in all these layers, we cannot understand the system and what it does. We also find the perspective of Stafford Beer useful (Beer, 2002), who, as a cybernetician, said that 'the purpose of a system is what it does', and not what it claims to do. What AI claims to do is readily visible from the literature produced by its developers and vendors. What it does, is more complex and requires situated research. If we compare research on AI governance to AI-based systems, we could say that most disciplines are using epistemologies that equate to symbolic reasoning – using hard-coded rules for what is relevant to the analysis, for choosing how to analyse that information, and for claiming that the resulting conclusions are valid. Critical data studies, like anthropology or some forms of philosophy, instead works in iterative interaction with the field: it proceeds through intuition and interaction, clustering inputs, seeking new patterns and correlations, and derives its validity from its underlying theory and methodological underpinnings, rather than by a stepwise explanation of process.

AI and ChatGPT 3.5

Prompt: Please explain what AI is, in 200 words, for an academic audience.

ChatGPT: AI refers to the ability of machines to simulate human intelligence and perform tasks that traditionally require human cognition, such as reasoning, problem-solving, perception, and learning. AI technologies encompass a broad range of techniques, including ML, NLP, computer vision, and robotics.

AI has a long history, dating back to the early 20th century, but recent advances in ML and big data have led to significant breakthroughs in AI research and application. AI is now used in a variety of fields, including healthcare, finance, transportation, and entertainment.

One of the most promising areas of AI is ML, which involves training algorithms to learn from data and improve their performance over time. This approach has led to significant advancements in speech recognition, image recognition, and NLP. Another important area is robotics, which involves the design and development of intelligent machines that can interact with the physical world.

While AI has great potential for improving human life and addressing complex societal problems, it also raises ethical and social concerns, including the potential for job displacement, bias, privacy infringement, and misuse. As such, the responsible development and deployment of AI technologies requires careful consideration of their social and ethical implications.

Studying and teaching trust

Trust, data systems, and George

The design of data systems is guided by several canonical concepts (guiding principles which underpin the field) towards realizing the dreams of Leibniz, Boole, Babbage, and company sketched in my earlier remarks on Intelligence. I will use one of these principles, the so-called ACID properties of transactional data systems, to illustrate a primary understanding of the concept of Trust in typical mainstream data engineering work.

The ACID properties govern the norms, the practical expectations, around data systems which manage transactions, that is, the updating and maintenance of a data collection as it evolves over time. ACID is typically taught in introductory courses on data management systems in CS and DS degree programmes. A classic example: when you make a purchase with your bank card, your bank balance must be correctly updated in the bank's database records for your bank account. ACID is an acronym denoting the four core principles of managing such transactions, to maintain the data system as the single point of truth: atomicity, consistency, isolation, and durability. Regarding Trust, 'durability' here means that an update should persist durably on some stable storage mechanism (for example, a hard drive) at the end of a transaction.[21] This means that even in the case of disruptions such as a system failure, the transaction will not be undone. In essence, a database system should be trusted to never forget data (in the absence of complete loss of the storage mechanism and all of its backup copies, such as in the case of a fire destroying the hard drive).

In the broader social contexts in which data systems are deployed, perfect remembering is often pathological behaviour. Humans forget, and for good reasons: personal growth, forgiveness, mental health. Societies forgive and forget, as reflected in laws governing the expungement of criminal records of minors and laws governing the right to erasure in digital systems. As Viktor Mayer-Schönberger has highlighted in his work, data forgetting is fundamental in a wide variety of social contexts (Mayer-Schönberger, 2009). This tension between the norms of data systems and the norms of society demands significant further study, especially in the data management research community.

Trust, law, and Aviva

A trust-framed look at law could start with the argument that laws are put in place to enable persons and institutions to interact in relative safety. Assuming that the need to trust arises when such safety cannot be assumed, law has an interest in trust relations that are formed to this end. The idea is that law serves society with democratically negotiated, knowable, foreseeable, and enforceable rules that allow subjects to rely on each other's and on institutional respect for public values and public order. Where law sees a role for itself, it operates from a standpoint of distrust, one could say: it defines trust-*worthiness*, *good* faith, *legitimate* expectations in light of the existence of their opposites. When things turn out badly, subjects can seek legal remedies and turn to legal institutions whose job it is to deal with cases of abused trust. Law Bachelor students are, for example, asked to memorize judgments about how X was reasonable to trust Y to not/act in a certain way in context Z, and that it was therefore reasonable that Y was condemned for abusing their position.

Lawmaking is a political activity that resonates with the outcomes of normative negotiations on all these points. And since legal rules work with semi-open norms, decisions on who and what counts as trustworthy are also responsive to those who populate the legal institutions, and research institutions, that interpret and further develop law. For trust in law itself, and in law-based societal rule, this matters greatly. Persons and communities whose values and interests are underrecognized in law and politics are less safe in their societies, and as a result they are trained in a rich variety of *dis*-trust.[22]

The subject of 'AI' is relatable to the aforementioned in many ways so I will briefly zoom in on a particular theme, namely insightfulness. First, to be able to investigate, interpret, track, and criticize how law co-shapes the normative landscape, that is, to gauge law's trustworthiness, law itself needs to be a sufficiently insightful process, practice, and paradigm. Law's progress has always depended on the hard work and activism of those who reveal law's biases, and continues to depend on them in 'AI-infused times' (Williams et al, 2022). The use of 'inscrutable' and/or experimental digital technologies in law-based policy, decision-making, and the judiciary, stands to complicate this challenging work. Second, it matters what standards law sets with regard to what constitutes trustworthy 'AI' and what kind(s) of insightfulness of AI practices are needed to assess this.

As mentioned in the 'intelligence' lemma above, in my teaching on the relations of insightfulness and trust in 'AI' contexts I like to discuss the longevity of the technological complexity argument. The argument was used in medicine around (and against) the paradigm shift from 'doctor knows best' to informed consent, that is, from trusted to trustworthy practices. Physicians who were uncomfortable with the shift at the time argued how medical knowledge was 'too technically complex' to ever be usefully understood by patients, and that trust should inevitably be invested in their medical authority. As Katz famously argued, however, other reasons for their explanation averseness hid behind the argument. Among other things he cites strong hierarchical cultures resulting in blind loyalty to medical elders; a medical lack of understanding of bodies and disease (especially for certain groups), medical chauvinism, and a widespread refusal to recognize the interests of patients (Katz, 1984). The arguments were supported by several equally problematic premises: that doctors can understand their patients well enough without engaging with their experience and standpoints, that 'medical knowledge' can be identified in isolation from other forms of knowledge, and that knowledge and decision-making about patients can proceed responsibly without patients' meaningfully informed participation in it. All these arguments and premises are recognizable in 'tech complexity' discussions about AI today, and in the technological lobby to keep law out of technological expertise-making while simultaneously calling for law to set up systems for accountability.[23]

To come back to the point of how groups' values and interests can be served better than those of others by law, attention should be called to how legal protections from wrongs and harms such as racism, marginalization, and discrimination have tended to fail less privileged communities well before 'AI'-driven decision-making became a widespread practice. One factor in why this is so is that law typically demands of individuals to prove how they were wronged in comparison with their peers, which is not possible when the abuse is systemic. In light of how 'AI' has been shown to exacerbate harms of this kind on a large scale, a progressive legal move could be to reverse burdens of evidence in contexts where subjects have historical reason to distrust the persons, institutions, and technologies they (need to) interact with. Legal efforts towards 'trustworthy AI' that disregard law's failures of protecting people from what are by now canonical 'AI' harms, can be considered a canonical failure.

Trust, political philosophy, and Gijs

Trust is not a canonical concept I use in general in my academic work, nor specifically in my lectures.[24] If I, however, was pressured to answer the question of the role of trust in political philosophy/theory, my answer would be something along the following lines. One strand of research in political theory/philosophy that might be more inclined to incorporate trust into their research agenda is research into the relationships between platforms, media, and the public sphere. For researchers working on these topics, democracy is often understood in a deliberative and knowledge-oriented endeavour where citizens, on the basis of publicly available information, discuss matters of public concern, as a means to reach agreement on how to deal with them. How information is shared with citizens, how the media works, how democratic debates and discussions are organized, and how truth is being produced in these processes are, therefore, key research topics. As part of such research on media and democracy, citizen's trust in both the facts they receive via the media, the media themselves, and each other, are topics to conduct research on.[25] Questions of trust understood from this angle, have received a significant boost in attention after the Cambridge Analytica scandal that showed how easy it is for big tech companies, through the manipulation of platforms/media, to affect the political decision-making process.[26] Trust, in other words, can become a topic of concern for political philosophers interested in technology but is then often considered to be part of a bigger puzzle on the functioning and quality of our democratic 'public spheres'. Scholars working in different subfields, or scholars like me who are not sure what to think of the ideal of the public sphere in the first place,[27] do not necessarily engage with 'trust'.

Trust, computer science, and Alexander

In computing, the topic of trust is predominantly discussed in the context of computer security: for example, a trusted computing base is a collection of hardware and software components deemed to be crucial for security of the entire system (Department of Defense, 1985), while a 'root of trust' is a 'tamper-resistant element in a digital system that can always be trusted, and therefore can be depended on as the root of all trusted operations' (Rowland and Karch, 2022). This interpretation of trust suggests that trust is being seen as a property of a system; it is aligned with the notion of trustworthiness – components that have met criteria related to correctness, safety, quality of service, security, and privacy have been deemed trustworthy (Hasselbring and Reussner, 2006). Alternatively, trust is associated with a reliable party certifying trustworthiness of the software: for example, macOS High Sierra *Trust* Store contains *trusted* root certificates,[28] issued by certificate authorities trusted by Apple, and subsequently by IT administrators of macOS High Sierra.

Another line of thought in computer security sees trust as a relation between communicating components (Garlan et al, 2009; Gai et al, 2010; also see more recent discussion of this topic in the context of the blockchain technology in Bryant and Saiedian, 2022) rather than property of an individual component; components then are seen as representing humans and performing series of simple actions on their behalf such as sending and receiving messages or performing computations. An example of this line of thought is the zero-trust security popularized by Google (Ward and Beyer, 2014), that, for example, states that 'a device that has not been updated with a recent OS patch level might be relegated to a reduced level of trust'. While presence of multiple often incompatible definitions of trust within the security community has been recognized as problematic, the common denominator seems to be that trust is absolute, established through complex hardware or cryptographic protocols and conflated with them.

Human–Computer Interaction (HCI) and, more recently SE, consider trust as a relation necessarily involving humans: for example, HCI positions itself in the space between individuals and software they interact with, while SE tends to take perspective of management science and study trust between members of software development teams. Finally, security aspects and trust as an interpersonal relation come together in security-related studies of topics related to deception and fraud in computer-mediated communication, for example, phishing.

Propagation of the AI techniques calls for rethinking the notion of trust and trustworthiness, triggering attention of researchers, practitioners, and policy makers to the topic of trustworthy AI, requiring AI to be lawful, ethical, and robust 'from a technical perspective while taking into account its social environment'.[29] In particular, robustness requires rethinking the aforementioned correctness, safety, quality of service, security, and privacy. The notion of correctness is particularly interesting: how can one distinguish inherent imprecision of AI techniques from bugs introduced in their implementation? Moreover, as more data or better AI techniques become available, answers produced by those techniques might change, requiring redefinition of the notion of correctness, trustworthiness, and ultimately truth.

Trust, data science, and Akrati[30]

In data science, when we talk about trust, it mainly refers to whether you can trust the output of the algorithm or an automated system. It works both ways, for scientists/researchers, it means to design systems that they can trust, and at the same time can be trusted by the users. To design trustable systems, data scientists consider trust at several steps, including:

- *Trustable data*: For computer scientists, trust in data is a critical consideration. Sometimes the data are generated due to biases in the system and cannot be used to train an AI model. For example, job hiring AI systems (Lewis, 2018) might have bias for female candidates. Beside this, data might be incomplete or manipulated (either intentionally or unintentionally), and this can have significant consequences. Therefore, computer scientists need to ensure that the data they are working with are accurate, reliable, and trustworthy.
- *Trustable algorithms*: In AI, we design algorithms/models to make decisions or recommendations, and in these cases, it is important to ensure that the algorithms are fair, transparent, and unbiased (Kleinberg et al, 2018). This means testing the algorithms rigorously and verifying that they are producing results that are consistent with expectations.
- *Trust in security*: Data scientists use databases to store the data and it is important to ensure their security from attacks. If the data storage is not secured, then it might risk users' data and they will not trust this system. Security measures can include things such as encryption, authentication, and access controls.
- *Trust in technology*: Finally, computer scientists need to consider trust in the technology itself. This means ensuring that the technology is reliable, robust, and performs as expected. It also means considering the ethical implications of the technology and ensuring that it is being used in a responsible and ethical manner.

By ensuring trust in these steps, data scientists can help ensure that their systems are trustable, reliable, accurate, and beneficial to society. However, even if a trustable system is designed, the main question is how to convince users that the system is trustable. For example, MYCIN was developed in the 1970s to assist medical professionals in diagnosing bacterial infections and recommending appropriate antibiotics to people. The success rate of MYCIN for prescribing correct medication was 65 per cent, that was higher than physicians who had an average success rate of 55 per cent (Liang, 1988). Despite its impressive accuracy and potential usefulness, people were hesitant to fully trust MYCIN for several reasons, including lack of transparency, limited knowledge base, and psychological factors.

The main question for data scientists is why people do not trust AI and how we can make people trust it. For the former question, the reasons are that there are several examples where AI has failed to qualify to be trustable, such as the hiring system (Raghavan et al, 2020), health care (Parikh et al, 2019), criminal identification (Angwin et al, 2016; Snow, 2018), facial recognition (Inioluwa and Buolamwini, 2019), translation systems (Prates et al, 2020), and so on. For later questions, AI scientists should focus on designing explainable systems with accountability, transparency, and ethical considerations. Beside these, educating people about AI systems and involving stakeholders in the design process might help people trust AI systems.

Trust, critical data studies, and Linnet

For critical data studies, the use of the terminology of 'trust' is a reliable sign that something is getting interestingly out of control. This is particularly true when the use of the term is purposely kept 'vague on who is to trust and whom, or what, is to be trusted'.[31] Keyes (2021) argues that current conceptualizations of trust in technology governance assume it is simple, transactional, and can be created to fit the needs of technology firms, and conversely that mistrust is defined as a flaw in users which must be fixed through rational explanation.[32] As such, it seems to be a continuation of the 'deficit model' of science communication, where people need to be improved and modernized so that they become more rational and appreciative of what science is doing (Simis et al, 2016). Trust, in this framing, can be understood as a companion to legal certainty for businesses: something that must be engineered and standardized so that innovation can proceed.

Conversely, there are more complex forms of trust reflecting subjective experience that are also interesting for scholarship of technology and its governance. Trust, for instance, as a dynamic component of relations between people and powerful entities such as corporations and public institutions. Here, trust in technology is functionally a subset of trust in power: people will trust to the extent that their interests are represented, and to the extent that the institutions they are trusting are accountable to them. Just as there would be no love poetry if love were always requited, trust would not enjoy the place it does in technology governance discourses if technological power were generally used in unproblematic ways. Therefore, as with democratic politics, the value of trust is frequently in the ways in which it is withheld.

'Trust' is most useful to critical data studies as a rhetorical signal that someone is in a position to behave badly at scale. When we hear the language of trust from regulators, rather than philosophers, we should ask what we are being asked to look away from. This is because most of the architectures that exist to prevent bad governance – electoral politics, the regulation of markets and businesses, research integrity codes, and most of public law, are designed to prevent people from having to trust the powerful. Critical data studies begins from the assumption that it is unlikely that technology is going to be the one phenomenon where this logic does not apply.

Trust and ChatGPT

Prompt: Please explain what trust is, in 200 words, for an academic audience.

ChatGPT: Trust is a fundamental concept in social sciences and humanities, and it is generally defined as the belief that an individual, group, or institution will behave in a reliable and predictable manner. Trust is a complex phenomenon that involves cognitive, emotional, and behavioural components, and it can be studied at different levels of analysis, from individual relationships to entire societies.

In interpersonal relationships, trust is often built through repeated interactions that demonstrate consistency, honesty, and integrity. When individuals trust each other, they are more likely to engage in cooperative behaviours and take risks together. However, trust can also be easily broken by betrayals, deception, or incompetence, and its restoration often requires significant effort and time.

In organizational contexts, trust plays a crucial role in shaping the behaviour of employees, customers, and stakeholders. Organizations that are perceived as trustworthy are more likely to attract and retain talent, build long-term relationships with customers, and generate positive reputations. Trust is also important in the context of institutions, such as governments and legal systems, which rely on the trust of citizens to maintain legitimacy and social order.

Trust has been studied from different theoretical perspectives, including social exchange theory, institutional theory, and social identity theory, and its mechanisms and consequences continue to be a subject of active research and debate in social sciences and humanities.

Discussion

The roughly formulated meta-questions with which we started this conversation were concerned with the relationships between our disciplinary backgrounds, the canonical concepts we choose, and the potential frictions emerging from the interaction of these different understandings, meanings, and worlds. At least three different themes pop up that relate to these questions, and that are worth explicating here.

On disciplinary straitjackets, and educating (in) DS

Many years ago, already, in a seminar series conveniently titled 'Discipline and Place' that one of us attended,[33] someone described academic disciplines as 'comfortable straitjackets'. Our discussion shows and confirms that the straitjackets we are all stuck in, determine how we (a) treat and (b) understand the canonical terms we chose to discuss. For legal scholars, for instance, 'intelligence' is directly related to human agency for reasons that have to do with the inherently normative character of the discipline of law. For data scientists, the concept of 'trust' is often understood as a response to the output of algorithmic or automated systems. It can even be an input to computing systems, rather than an output, because it can be defined as a quality possessed by infrastructures which are then used for computing processes. Researchers in critical data studies, by contrast, see the concept as a sign that something is going horribly wrong.

The roles of boundary objects

In contrast to discussing a particular disciplinary *understanding* of a canonical concept (b), some of us limited themselves to a description of how a term is treated in their discipline (a), without an accompanying substantive understanding of it. For political philosophers, concepts like algorithms, technology, and AI are often used interchangeably and primarily for the purpose of discussing several new and important moral-political problems the bad usage of such technologies could result in. In more traditional computer science curricula, AI is treated similarly: rather than presenting a specific analysis of the concept, it is being used as an umbrella term that incorporates all sorts of topics that in some sense relate to 'AI'. On the one hand, these descriptions might amount to a different way of *defining* and *understanding* things from within the perspective of that discipline ('one describes rather than defines'). On the other hand, and we take this to be more plausible, can these *treatment*s of especially 'AI' be interpreted as the instantiation of a boundary object ('AI') that fulfils various disciplinary functions. It allows philosophers, for instance, to connect themselves to new problems relating to 'technology' and by doing so further their field. Or, which is also not unimaginable, it helps to secure funding for research into data and technology-related research. What is key here, is that these usages of AI do not necessarily have to be accompanied with a full-fledged definition, conception, or understanding of 'AI'. It is all pleasantly fuzzy.

Tensions between disciplines, boundary objects, and education

Next to productive interactions resulting from the construal of boundary objects situated in between the computational fields and disciplines and the more social scientific and humanities ones, we did also notice destructive tendencies related to our situatedness in disciplines, and the need to connect to others. Such tensions manifest across a wide range of academic activities related to the way we build our fields, for instance when we found journals, publish, organize conferences, teach, or form research groups and institutes. To look at a few of these in more detail: when we seek support for our work, we are usually expected to do so under the auspices of a particular discipline, which determines how funders will read our proposals. In what is often a check-box exercise for reviewers, interdisciplinarity (where it is a criterion) must be identifiable, measurable, and calculable. Conversely, so must disciplinary affiliation. In the EU's Horizon grant programmes involving technology, for instance, there is often tacit signalling as to which disciplines are expected to participate in a particular call, and any proposal involving an unexpected mix of disciplines or the placement of a given researcher in an unconventional role must be clearly flagged and justified, and will usually raise scepticism from reviewers.

Conferences focusing on boundary objects such as 'AI' are becoming to some degree interdisciplinary, perhaps most of all the ACM's FAccT series on fairness, accountability, and transparency in computing. These latter series of conferences, however, are characterized by problems of different understandings of core concepts such as power, ethics, and bias (Ganesh et al, 2020). Moreover, they surface different disciplinary understandings of what constitutes ethical funding practices, with the norm for computing sciences being close collaboration with industry due to their predominantly applied nature and fluid boundaries with commercial technology providers, and the norm for social science and humanities disciplines being to keep much more distance from it for fear of ideological capture (Young et al, 2022). For the computing sciences, keeping up with developments in, and the needs of, industry, is a mark of good research. For the social sciences and humanities, influence by industry is closely controlled through research ethics codes, and ties with industry actors are scrutinized (though paradoxically also increasingly encouraged due to funding constraints) by universities.

Teaching across disciplines also surfaces these frictions. Students in DS are often, though not always, on a path towards careers in industry, and courses are designed to prioritize applied knowledge. The requirement to follow a course in ethics while doing a degree in DS is frequently seen as introducing an alien way of thinking into an otherwise coherent degree programme, where students being trained to think in a positivist paradigm must suddenly learn, often for the first time, to engage with critical scholarship on technology. Empirically, teachers have found that the introduction of deontological modes of reasoning about ethics, in particular, is hard for many students from an exact science background, who tend to find a consequentialist (even utilitarian) perspective most intuitive (Taylor and Dencik, 2020).[34] This is perhaps one of the most significant, but almost entirely unscrutinized, fault-lines between computing sciences and the social sciences and humanities: the assumption of a consequentialist rationale for both research goals and the application of technology is such a basic underpinning for the computing sciences that it is difficult for students to imagine their way out of it – and for meaningful forms of ethical scrutiny to be applied to computing and data science projects (Metcalf and Crawford, 2016).

On (de)professionalizing boundary objects

Our contributions suggest that the incorporation of canonical concepts in disciplines, as a means to develop a discipline internally or to connect it to other fields, disciplines, or funding streams, should not be left uninterrogated.

For starters, as highlighted in different ways by several of us, the incorporation of generalized concepts without careful scrutiny – a standard practice across our fields – could result in a limited grasp of the problems researchers should deal with. Platforms, for instance, cannot be equated with algorithms, which is the reason critical data studies researchers deconstruct and expose the various layers that together make 'AI'.

Going slightly beyond such epistemologically related limitations, we also notice political economic worries with respect to the treatment of these and other canonical concepts in our disciplines. The pragmatic cross-disciplinary agreements on boundary objects we have explored earlier have the effect of making it possible to stabilize, and therefore also professionalize, the process of studying, teaching, and monetizing them. There has to be some coherence in our DS programme in how we use concepts, because, among other things, one has to find consensus on definitions when applying for one's interdisciplinary grant proposals. As such, we might see technological boundary objects in the world of data and AI *becoming* boundary objects through their links to industry and public and private funders. The tech-oriented boundary objects stabilize and reinforce relationships and collaborations between education and industry to such an extent that our capacities to come up with, and interject in our classes, objects and languages that are oriented towards other-than-industrial languages and worlds has been reduced.

Disciplinary affiliation, and the security offered by these straitjackets, has a price, and attempts to critically interrogate the fundamental assumptions of our disciplines puts us out of step with colleagues with whom we could otherwise be collaborating on papers, education, and grant proposals.

An important characteristic of the boundary objects encountered in our canons, in sum, is the incentivization of disciplinary obedience geared towards the production of efficient and frictionless collaboration. By taking our canons as the starting point of our conversation, we acquired a better understanding of the boundary objects that help to communicate without really understanding one another. The resulting lessening of the comfort granted by our straitjackets will, we hope, increase the room we have for manoeuvre within and in between our fields, potentially resulting in less efficient, productive, and frictionless possibilities not directed towards disciplinary renewal, but a better world.

Appendix: canonical terms
- Trust
- Algorithms
- Governance
- Ethics
- Explanation
- Law
- Infrastructure
- Accountability
- Data
- Fairness
- Justice
- Causality
- Truth
- Efficiency
- Significance
- Ontology
- AI
- Politics
- Power
- Ambiguity
- Accuracy
- Probability

Notes
[1] Tilburg University, the Netherlands, aviva.degroot@tilburguniversity.edu
[2] Eindhoven University of Technology, the Netherlands, g.h.l.fletcher@tue.nl
[3] Tilburg University, the Netherlands, g.vanmaanen@tilburguniversity.edu

4   Leiden University, the Netherlands, a.saxena@liacs.leidenuniv.nl
5   Eindhoven University of Technology, the Netherlands
6   Tilburg University, the Netherlands, l.e.m.taylor@tilburguniversity.edu
7   I write from the perspective of a data language researcher, trained and working for the past two decades in the domain of data management systems, a subfield of the computing and data sciences. I have been on the faculty at a Dutch technical university for the past 14 years; prior to this I was an assistant professor at a West-coast American comprehensive public university. Before completing my PhD on topics in data integration (at a comprehensive public university in the American Midwest), I double majored in mathematics and liberal studies in cognitive science, writing my Bachelor's thesis on topics in cognitive science viewed through a social philosophy lens (at a comprehensive public university in the American South). As a child I lived between Japan, with extended family centred on Yonaguni Island, and America, with extended family centred in the Appalachian mountains of western North Carolina. I teach in Bachelor's and Master's programmes in computing and data science.
8   For a recent discussion of examples of these harmful aspects and their historical roots, see Whittaker (2023).
9   I am a postdoc in AI & Human Rights, a position I landed after a PhD in which I dissected several canonical legal 'explanation obligations' in light of how such obligations are seen to be hard to meet in the context of today's automated decision-making methods. Informed by historical accounts of both AI and law, and by earlier work experience in the field of legal aid, I am increasingly sceptical of such 'novelty' leaning problematizations. Working with differently trained researchers helps me to identify new iterations of old problems, and to think through how solutions can take a more holistic approach. In doing so I am confronted with how the separation of types of work, types of workers, and the power relations that attach to these separations poses obstacles for more inclusive work towards (global) social justice. But working this way can also complicate single-discipline work collaboration. Working with more singularly oriented legal scholars, I tend to (feel the need to) unsettle disciplinary understandings of, for example, 'well-established' fundamental values. Whether that happens also depends on colleagues' more personal standpoints. For example, differences in legal focuses and standpoints also tend to express in, and as, political colour, which influences the type of students a university attracts and the scholarship that is produced by its researchers. Adding more personal bios to articles seems like a good idea in light of this, especially in collective endeavours with one person per discipline. Aviva, 'the' lawyer? No, Aviva, 'a' lawyer. I have professional backgrounds in cabinet making, (SF and horror) filmmaking, and legal aid – where protecting persons against an ill-wishing, tech-happy State was core business.
10 https://edri.org/what-we-do/decolonising-digital-rights/
11 I studied history and (political) philosophy, and focused on democratic theory, migration/multiculturalism, feminist/new materialist theories, and science and technology studies (STS). When looking for a PhD position I ended up at Tilburg Law School where I joined an administrative law project on 'citizen-friendly data communication', and where it was my job to reflect on ethical principles that could help to guide such communicative processes. Being uncomfortable with doing 'ethics' in that way, I decided to do an empirical analysis of open government and open data policies/practices in the Netherlands instead. Methodologically speaking, this amounted to a mixture of STS/ethnography and political theory. Afterward, during my postdoc at a private law institute elsewhere, I combined literature from political theory and (political) economics in a research project on collective data governance. I, in other words, consider myself to be a disciplinary migrant that travels to and tries to make sense of the academic debates and disciplines that help to make sense of a problem. I teach in the Bachelor's Data Science progamme in Tilburg/Eindhoven where

I try to make the students receptive to ethical-political problems related to 'technology', and draw from a mixture of political philosophy/theory, data ethics (broadly conceived), critical data studies, and STS. For this contribution, I primarily thought about how from a political philosophical point of view, the two canonical concepts would be approached, and drew from more accessible/introductory texts on the matter (of which there are few).

12 Though, for example, Risse in the beginning of his book distinguishes between algorithms, machine learning algorithms, 'specialised AI', and 'general AI', the concepts are often used relatively interchangeably in the rest of the book (Risse, 2023).

13 Coeckelbergh wrote one of the few accessible overviews of the relationship between political philosophy/theory and AI/algorithms, and prioritized the problems that different technologies such as recommender systems pose to, for instance, human freedom (Coeckelbergh, 2022).

14 See both Risse (2023) and Coeckelbergh (2022).

15 Drawing from critical/Marxist work on the 'commons', Thijs Lijster makes this argument (Lijster, 2022). Ugur Aytac also presented a similar argument at the Mancept workshops 2022: https://sites.manchester.ac.uk/mancept/mancept-workshops/programme-2022-panels/digitaldemocracy/

16 See, for example, the work of Anne Helmond: www.annehelmond.nl/publications/

17 As Boris Johnson labelled the Ofqual algorithm used to predict the grades of students that were unable to do their final exams due to the COVID pandemic: www.bbc.com/news/education-53923279

18 I am a classically trained computer scientist that has slowly migrated to software engineering and, specifically, to studies of human and social aspects of software engineering. While my expertise is in software engineering, in the texts, I am trying to present a broader view of CS in general. This is, of course, a challenging endeavour due to the variety of topics studied in CS and different perspectives CS disciplines have on the topics discussed in this chapter.

19 I am a computer science engineer by training and then extensively worked on network science and data science. I write from the perspective of a computer scientist and especially, a data scientist. Currently, I am working as an assistant professor at Leiden Institute of Advanced Computer Science, Leiden University, the Netherlands. I have taught data mining and related courses to Bachelor's and Master's students and written about how students learn AI and then what AI means to them.

20 I am a human geographer and critical data studies researcher working in the law faculty at Tilburg University. I have taught in International Development Studies, Data Science, and international governance at Bachelor's and Master's levels.

21 The others stand for, respectively, atomicity, consistency, and isolation.

22 The extent to which lawmakers and legal institutions care about this depends. When a critical mass of legal subjects lose trust in rule-based society, law, and the powers that law is assumed to keep in check, both stand to lose legitimacy – and power.

23 See, for example, the call to establish licensing systems for building 'cutting edge models' and establishing liability for foreseeable and preventable harms: https://managing-ai-risks.com/

24 It is not extensively being discussed by either Risse or Coeckelbergh, and also the 'tech ethics' handbooks do not include it in their indexes (van der Poel and Royakkers, 2011; Taebi, 2021; Nyholm, 2023). An exception is a chapter on Kantian ethics in a recently published technology ethics anthology (Myskja, 2023).

25 For example, Farrell and Schwartzberg, 2021.

26 Coeckelbergh does discuss such problems in chapter 4 of his book ('Democracy: echo chambers and machine totalitarianism'), though does not give a substantive analysis of the role of trust in these matters.

[27] See, for example, the collection of reflections on Habermas' concept in Calhoun (1992).

[28] https://support.apple.com/en-us/HT208127

[29] https://digital-strategy.ec.europa.eu/en/library/ethics-guidelines-trustworthy-ai

[30] I am writing about what trust means for data scientists and for people using these AI systems.

[31] https://algorithmwatch.org/en/trustworthy-ai-is-not-an-appropriate-framework/

[32] https://reallifemag.com/standard-evasions/

[33] Gijs participated in the mentioned seminar. A course overview can be found here: https://studiegids.universiteitleiden.nl/en/courses/101290/discipline-and-place-in-the-social-sciences-and-the-humanities

[34] Consequentialism and utilitarianism are closely related strands in the philosophical literature – and especially, moral and political philosophy – that evaluate the (moral) rightness of our actions on the basis of the consequences they would have. See https://plato.stanford.edu/entries/consequentialism/. This can be contrasted with theories that prescribe that actions should be in accordance with particular moral rules or principles ('You should never lie, regardless of the consequences of your (in)action').

References

Ardis, M., Budgen, D., Hislop, G. W., Offutt, J., Sebern, M., and Visser, W. (2015) SE 2014: Curriculum guidelines for undergraduate degree programs in software engineering. *Computer*, 48(11): 106–109.

Agre, Philip. (1995) The soul gained and lost: artificial intelligence as a philosophical project. *Stanford Humanities Review*, 4(2): 1–19.

Agre, Philip. (1997) 'Toward a Critical Technical Practice: Lessons Learned in Trying to Reform AI'. In: Geoffrey C. Bowker, Susan Leigh Star, William Turner, and Les Gasser (eds) *Social Science, Technical Systems and Cooperative Work: Beyond the Great Divide*, 131–158. Hillsdale, NJ: Erlbaum.

Angwin, J., Larson, J., Mattu, S., and Kirchner, L. (2016) 'Machine bias'. ProPublica, 23 May. Available at: www.propublica.org/article/machine-bias-risk-assessments-in-criminal-sentencing

Beckwith, J. and Pierce, R. (2018) 'Genes and Human Behavior: Ethical Implications'. In: Robert T. Gerlai (ed) *Molecular-Genetic and Statistical Techniques for Behavioral and Neural Research*, 599–622. San Diego: Academic Press. https://doi.org/10.1016/B978-0-12-804078-2.00025-8.

Beer, S. (2002) 'What is cybernetics?' *Kybernetes*, 31(2): 209219.

Benjamin, R. (2019) *Race after Technology*. New York: Polity Press.

Bigenwald, A. and Chambon, V. (2019) 'Criminal responsibility and neuroscience: no revolution yet'. *Frontiers in Psychology*, 10: Article 1406. https://doi.org/10.3389/fpsyg.2019.01406

Bloom, H. (1994) *The Western Canon: The Books and School of the Ages*. New York: Harcourt Brace & Company.

Bridges, Khiara M. (2017) *The Poverty of Privacy Rights*. Stanford, CA: Stanford University Press.

Broussard, M. (2023) *More than a Glitch: Confronting Race, Gender, and Ability Bias in Tech*. Cambridge, MA: MIT Press.

Bryant, B. D. and Saiedian, H. (2022) Key challenges in security of IoT devices and securing them with the blockchain technology. *Security and Privacy*, 5(5).

Calhoun, C. (ed) (1992) *Habermas and the Public Sphere*. Cambridge, MA: MIT Press.

Coeckelbergh, M. (2022) *The Political Philosophy of AI: An Introduction*. Cambridge: Polity Press.

Crawford, K. and Joler, V. (2018) 'Anatomy of an AI system'. Available at: www.anatomyof.ai

D'Souza, R. (2018) *What's Wrong with Rights? Social Movements, Law and Liberal Imaginations*. London: Pluto Press.

de Groot, A. (2023) Care to explain? A critical epistemic in/justice based analysis of legal explanation obligations and ideals for 'AI'-infused times. PhD thesis. Tilburg University, Tilburg.

Deleuze, G. and Guattari, F. (1987) *A Thousand Plateaus: Capitalism and Schizophrenia*. Minneapolis: University of Minnesota Press.

Department of Defense (1985) 'Department of Defense trusted computer system evaluation criteria', DoD 5200.28-STD

Dunn, J. (2005) *Setting the People Free: The Story of Democracy*. London: Atlantic Books.

Eubanks, V. (2018) *Automating Inequality: How High-tech Tools Profile, Police, and Punish the Poor*. New York: St. Martin's Press.

Farrell, H. and Schwartzberg, M. (2021) 'The Democratic Consequences of the New Public Sphere'. In: Lucy Bernholz, Hélène Landemore, and Rob Reich (eds) *Digital Technology and Democratic Theory*, 191–218. Chicago, IL: University of Chicago Press. Chicago Scholarship Online. https://doi.org/10.7208/chicago/9780226748603.003.0008

Floridi, L. (2016) On human dignity as a foundation for the right to privacy. *Philosophy & Technology*, 29(4): 307–312. https://doi.org/10.1007/s13347-016-0220-8

Fuchs, C. (2014) Digital prosumption labour on social media in the context of the capitalist regime of time. *Time & Society*, 23(1): 97–123. https://doi.org/10.1177/0961463X13502117

Gadamer, H.-G. (2014) *Truth and Method*. London: Bloomsbury Academic.

Gai, X., Li, Y., Chen, Y., and Shen, C. (2010) 'Formal definitions for trust in trusted computing'. UIC/ATC Workshops 2010: 305–310.

Ganesh, M. I., Dechesne, F., and Waseem, Z. (2020) Two computer scientists and a cultural scientist get hit by a driver-less car: a method for situating knowledge in the cross-disciplinary study of FAT in machine learning: translation tutorial. *Proceedings of the 2020 Conference on Fairness, Accountability, and Transparency*, pp 707–707.

Garlan, D., Allen, R., and Ockerbloom, J. (2009) Architectural mismatch: why reuse is still so hard. *IEEE Software*, 26(4): 66–69.

Haraway, D. (1991) *Simians, Cyborgs and Women: The Reinvention of Nature.* New York: Routledge.

Harding, S. (1991) *Whose Science? Whose Knowledge? Thinking from Women's Lives.* New York: Cornell University Press.

Hasselbring, W. and Reussner, R. H. (2006) Toward trustworthy software systems. *Computer,* 39(4): 91–92.

Inioluwa, R. D. and Buolamwini, J. (2019) Actionable auditing: investigating the impact of publicly naming biased performance results of commercial AI products. *Proceedings of the 2019 AAAI/ACM Conference on AI, Ethics, and Society (AIES'19).* New York: Association for Computing Machinery, pp 429–435. https://doi.org/ 10.1145/3306618.3314244

Katz, J. (1984) *The Silent World of Doctor and Patient.* Baltimore, MD: Johns Hopkins University Press edition, 2002.

Katz, Y. (2020) *Artificial Whiteness: Politics and Ideology in Artificial Intelligence.* New York: Columbia University Press.

Keyes, O. (2021) 'Standard evasions'. *Real Life,* 30 August. Available at: https://reallifemag.com/standard-evasions/

Kleinberg, J., Ludwig, J., Mullainathan, S., and Rambachan, A. (2018) Algorithmic fairness. *AEA Papers and Proceedings,* 108: 22–27.

Lepore, J. (2020) *If Then: How the Simulmatics Corporation Invented the Future.* New York: Liveright Publishing.

Lewis, N. (2018) 'Will AI remove hiring bias?' *Strategic HR Review,* 11 November. Available at: www.shrm.org/topics-tools/news/talent-acqu isition/will-ai-remove-hiring-bias

Liang, T. P. (1988) Expert systems as decision aids: issues and strategies. *Journal of Information Systems,* 2(2): 41–50.

Lijster, T. (2022) *Wat We Gemeen Hebben: Een Filosofie van de Meenten.* Amsterdam: De Bezige Bij.

Lindsay, R. K., Buchanan, B. G., Feigenbaum, E. A., and Lederberg, J. (1993) DENDRAL: a case study of the first expert system for scientific hypothesis formation. *Artificial Intelligence,* 61(2): 209–261.

Mayer-Schönberger, V. (2009) *Delete: The Virtue of Forgetting in the Digital Age.* Princeton, NJ: Princeton University Press.

Metcalf, J. and Crawford, K. (2016) Where are human subjects in big data research? The emerging ethics divide. *Big Data & Society,* 3(1): 2053951716650211.

Mhlambi, S. (2020) From rationality to relationality: Ubuntu as an ethical and human rights framework for artificial intelligence governance. *Carr Center for Human Rights Policy Discussion Paper Series,* 9: 31.

Mol, A. (2014) *The Body Multiple: Ontology in Medical Practice.* Durham, NC; London: Duke University Press.

Moyn, S. (2019) *Not Enough: Human Rights in an Unequal World.* Cambridge, MA: Harvard University Press.

Myskja, Bjørn K. (2023) 'Technology and Trust – A Kantian Approach'. In: Gregory J. Robson and Jonathan Y. Tsou (eds) *Technology Ethics: A Philosophical Introduction and Readings*, 122–129. New York: Routledge.

Nyholm, S. (2023) *This is Technology Ethics: An Introduction*. Hoboken, NJ: Wiley.

Parikh, R. B., Teeple, S., and Navathe, A. S. (2019) Addressing bias in artificial intelligence in health care. *Jama*, 322(24): 2377–2378.

Poole, G. (2009) 'Academic disciplines: Homes or Barricades?' In: Carolin Kreber (ed) *The University and its Disciplines: Teaching and Learning Within and Beyond Disciplinary Boundaries*, 55–57. New York: Routledge.

Prates, M. O., Avelar, P. H., and Lamb, L. C. (2020) Assessing gender bias in machine translation: a case study with google translate. *Neural Computing and Applications*, 32: 6363–6381.

Raghavan, M., Barocas, S., Kleinberg, J., and Levy, K. (2020) Mitigating bias in algorithmic hiring: Evaluating claims and practices. *Proceedings of the 2020 Conference on Fairness, Accountability, and Transparency*, pp 469–481.

Risse, M. (2023) *Political Theory of the Digital Age: Where Artificial Intelligence Might Take Us*. Cambridge: Cambridge University Press.

Rowland, M. and Karch, B. (2022) 'A Review of Technologies that can Provide a 'Root of Trust' for Operational Technologies'. Sandia Report SAND2022–3755.

Sandvik, K. B. (2023) *Humanitarian Extractivism: The Digital Transformation of Aid*. Manchester University Press.

Séailles, G. (1904) 'L'enseignement secondaire et la philosophie'. In: *Éducation ou revolution*. Paris: Armand Colin.

Shortliffe, E. (ed) (2012) *Computer-based Medical Consultations: MYCIN* (Vol. 2). New York: Elsevier.

Simis, M. J., Madden, H., Cacciatore, M. A., and Yeo, S. K. (2016) The lure of rationality: why does the deficit model persist in science communication? *Public Understanding of Science*, 25(4): 400–414.

Snow, J. (2018) 'Amazon's face recognition falsely matched 28 members of Congress with mugshots'. American Civil Liberties Union. Available at: www.aclu.org/blog/privacytechnology/surveillance-technologies/amazonsface-recognition-falsely-matched-28

Spivak, G. C. (1990) The making of Americans, the teaching of English, and the future of culture studies. *New Literary History*, 21(4): 781–798.

Star, S. L. and Griesemer, J. R. (1989) Institutional ecology, 'translations' and boundary objects: amateurs and professionals in Berkeley's Museum of Vertebrate Zoology, 1907–39. *Social Studies of Science*, 19(3): 387–420. https://doi.org/10.1177/030631289019003001

Taebi, B. (2021) 'Ethics and Engineering: An Ethics–Up–Front Approach'. In *Ethics and Engineering: An Introduction* (Cambridge Applied Ethics), 1–20. Cambridge: Cambridge University Press.

Taylor, L. and Dencik, L. (2020) Constructing commercial data ethics. *Technology and Regulation*, 1–10.

Taylor, L., van Maanen, G., de Ridder, J., Taebi, B., and Bruijnincx, P. (2023) Space to think. An analysis of structural threats to academic freedom and integrity. Amsterdam, KNAW. Available at: www.dejongeakademie.nl/publicaties/2495595.aspx

Tierra Común (2021) *Interventions for data decolonization*. Available at: www.tierracomun.net/en/home

van der Poel, I. and Royakkers, L. (2011) *Ethics, Technology, and Engineering: An Introduction*. Hoboken, NJ: John Wiley.

Ward, R. and Beyer, B. (2014) 'BeyondCorp: a new approach to enterprise security'. login Usenix Mag., 39(6).

Whittaker, M. (2023) Origin stories: plantations, computers, and industrial control'. *Logic(s)*, Issue 19.

Williams, P., Kuntsman, A., Nwankwo, E., and Campbell, D. (2002) 'Surfacing systematic (in)justices: a community view'. Available at: https://systemic justice.ngo/wp-content/uploads/2023/06/SystemicJustice_Report_FINAL.pdf

Wright, S. (2001) *International Human Rights, Decolonisation and Globalisation Becoming Human*. London: Routledge.

Young, M., Katell, M., and Krafft, P. M. (2022) 'Confronting power and corporate capture at the FAccT conference'. *Proceedings of the 2022 ACM Conference on Fairness, Accountability, and Transparency*, pp 1375–1386.

Reflection: 'Spilling the Authorship Tea' – On Herding Cats and Collective Response-ability

Jo Bates and Juliane Jarke

Introduction

The aim of this book was to facilitate dialogue between scholars in critical data studies across different disciplines, locations, and levels of seniority. This presents a collective (and alternative) approach to knowledge-making and dissemination to more standard contemporary models (Peters et al, 2016; Jandrić et al, 2023). It is, however, also a 'continuous struggle for meaning-making' (Jandrić et al, 2023: 249). For example, in reflecting on their experience of collective writing, Jandrić et al report that it 'took a leap of faith' when they contributed to their first collective paper. They also had so many questions: 'What does it mean to write together? How can we combine people's diverse ideas and strands of thinking into a coherent whole?'

We (Juliane and Jo) began this book project with similar questions. Leading a book project that kicked off with 78 academic authors across 50 author teams, of varying years' experience based in 14 countries across different time zones was unsurprisingly challenging! When we met up with our editor Paul at STS Graz in early May 2023, as we got towards the end of our initial round of workshops and authors were working on their chapters, Paul asked us 'Would you do it again?'. It was a difficult question to answer at that stage, things had gone relatively smoothly so far, but we were still unsure what was in store for us as the chapters came together. Now as the project draws to a close and we reflect on our work on it over the last 12 months, Jo overall has been pleasantly surprised it was not as difficult as it might have been to get all the chapters to the finish line, and would consider doing it again (keeping in mind some important lessons learned). Juliane's curiosity to explore how such an approach would advance dialogues and engagement in critical data studies was her initial starting point. At times she felt that she underestimated the work of 'herding cats' and 'building narratives' (Jandrić et al, 2023), but realized in the end the great potential of such an approach for collective knowledge-making and for building stronger ties among critical data studies scholars. In this chapter, Jo Bates and Juliane Jarke, the editors of this book, report and reflect on the collective writing process of this book together with some of its authors.

The collective writing process

Data Power was organized as a three-day in-person and online international conference in June 2022 by Data Power teams at The University of Sheffield in the UK, Carleton University in Canada, and the University of Bremen in Germany. Overall, 175 participants presented 122 research papers. In the months following the conference, we began to make plans for a book project emerging from the conference that would further support and enable dialogue across locations, disciplines, and career levels. Juliane had had an early conversation with Paul Stevens, our editor at Bristol University Press, about her ideas for a collaboratively authored book project which he was excited to support. From there Juliane brought the idea to the Data Power team, and convinced Jo to join the project as a co-editor.

We kicked off the project by exploring possible chapter themes with Tracey Lauriault, who had hosted the Ottawa stream of the 2022 conference. Some of these themes were based directly on the panels that were constructed for the conference, others were new formations of contributors it was possible to bring together, freed from the constraints of time zones and modes of participation as we were when constructing the conference programme. The 'Environmental Data Power' chapter was one such theme. This process of crafting emergent themes from the conference was clearly a subjective process, driven by our own insights and desires for where we would like to foster new conversations and collaboration on pressing topics. Once we had a list of thematic chapters and potential authors for each, we split the chapters between the two editors so we were leading on four each. In addition, we invited the keynote panel from Bremen, who had already engaged in an interdisciplinary dialogue for a conference keynote, to contribute a chapter on their approach of collective sensemaking and dialogue across data science, software engineering, law, philosophy, and science and technology studies. In November 2022, five months after the conference, we began sending invitations out to selected authors – around 5–8 per chapter. Given the novel approach of the writing project, we were unsure how people would respond, however, we soon received positive responses from most of the invitees, and began scheduling dates for chapter workshops in the new year.

Our first workshop was held in February 2023. Every participant was asked to prepare a 5 to 7 minute provocation in relation to the theme of the workshop in order to introduce themselves to the others and embark on a first exploration and mapping of the different perspectives that participants brought to the theme. We gave a brief introduction to the idea of the book and how we envisaged the writing process would take shape. This first workshop went well and so we kept the format for the kick-off workshops with the other chapters. The remaining chapter workshops took place in March and April. In all workshops, the workshop participants were asked to introduce themselves and their research through a short provocation. In order to do so, some used their slides from the Data Power Conference, others spoke about what they were interested in or concerned about in relation to the theme we set out. It quickly became clear that each author team had its own dynamics. What we learned from running one workshop was not necessarily useful in the next. We found that depending on the thematic focus of the chapter, participants seemed to find it easier or more difficult to pin down or frame a common ground (and some boundaries) for their endeavour. Some groups developed an argument and structure for their chapter with ease, outlining where each section/perspective of individual authors would go; others found that quite challenging. Some groups took up on our initial title of the chapter; others challenged our framing and developed it further. Some groups fairly seamlessly evolved their own collaborative approach, while others needed more steering. Some groups fostered a somewhat non-hierarchical approach, others had more dominant personalities who had more to say about the direction of travel. Groups also differed with respect to how they approached their chapter, for example, was it a good starting point to consider the most relevant concepts and controversies or was it better to think about what the authors wanted the readers to take away. These seemingly subtle differences had an impact on how 'dialogical' the chapters were written. Our role oscillated between being a moderator in a dialogue or panel and being an editor that needed to help develop a written (and for others sensible) account of the dialogue.

Despite the different cultural dynamics of each author collective, over the coming months they all made headway in writing their chapters. While follow-up workshops were held for some groups, that was challenging for some chapters where authors had busy schedules and were spread across different time zones, so authors often worked asynchronously on chapters using Google Docs, usually with separate documents for draft chapters and meeting notes. Each author contributed a section to the paper, and volunteers from the team were sought to contribute to framing the chapter in the introduction and conclusion. While there was a somewhat standard length for each individual section, author teams then decided their own approach to constructing an introduction and conclusion that worked for the contribution they were trying to make. Some also decided that they wanted to add positionality statements to their chapters. Discussions also took place to decide the order of the sections and how this contributed to the overarching narrative. As with the other components of the chapters, this was easier for some groups than others with some needing more editorial support than others.

As the chapters were taking shape, with input from author teams, we began drawing up shortlists of discussants for each chapter. Initially, we had planned to do these sections as transparent chapter reviews published after each chapter, but after some discussion with the publishers we opted for a 'safer' option of inviting discussants to contribute a section, taking on a similar role to a discussant on a conference panel. Overall, what was important to us was that we would further open the dialogue and involve scholars from outside the Data Power Conference community. We aimed for a mix of leading academics in the field and earlier career researchers to fulfil these roles, with 1–2 discussants per chapter. Over summer 2023, chapters were sent to selected discussants. When the responses came back, many of the author collectives appreciated the (critical) engagement with their chapter. Some decided to provide a response to the discussant sections either as individual contributors or collective statements, others were happy for the discussant to conclude the dialogue.

Specific challenges we encountered, and which we think are likely to crop up if we – or others – ran a similar project in the future, included the variability in commitment that people were able to offer, which impacted some people's capacity to keep to deadlines or volunteer for the extra work of contributing to a chapter introduction or conclusion once their section was complete. Across the chapters a small number of researchers dropped out due to over commitment and personal issues, sometimes resulting in a leadership vacuum in a chapter, meaning more input for editors. Most of these people communicated openly about their challenges, keeping us in the loop; others disappeared leaving us uncertain about their contribution. Personality and power dynamics also came into play. For example, some authors were reluctant to comment on one another's sections, while others were keen to do so, and there were sometimes group dynamics to manage as established academics took the lead in shaping the narrative of the chapter in line with their own perspective. As discussed above, different author groups took different approaches to the collaboration, and in some cases where the approach was more individualized this made it challenging to overcome a level of fragmentation in the narrative of the final chapter.

Despite the challenges, what worked well was the sense of collegiality and connection that emerged to varying degrees around each chapter. From the initial workshop through multiple emails and comments in Google Docs, connections were built that would not have happened otherwise, including between early career researchers and more established academics in the field. Being in the critical data studies field can mean you are a border hopper – with feet in different disciplinary spaces, so there is not always the opportunity to build these connections as one might in a more traditional disciplinary formation. The writing of the book hopefully provided a space for this for some of our contributors.

So, on reflection what do we think is the glue that makes such a process successful despite the challenges? Through our discussions we came to the conclusion that trust and collective response-ability were at the core of our process, from those initial moments when Juliane suggested the project, and Jo – despite feeling uncertain about what it would entail – trusted in her vision of collective knowledge production and dialogic engagement, through to the final product in which we trusted the readers to be open to what was presented to them. As Linnet Taylor comments in Chapter 9, 'For CDS, the use of the terminology of "trust" is a reliable sign that something is getting interestingly out of control', and we would agree that for a project such as this to work some element of control has to be lost, with trust – in the people, process, and vision – taking its place. As editors we trusted in one another, as well as the invited authors and discussants, to stay committed, communicative, constructive, and collegial. We also trusted in the process of collective writing to deliver tangible, engaging results. For everyone else that was involved there were likely also moments of uncertainty and a sense of things being 'interestingly out of control', when they leaned on their trust in their author and editorial teams to produce the chapters. This seems to be a general insight into collective writing as Jandrić et al (2023) similarly reflect on a 'leap of faith' that was necessary when embarking on such an endeavour.

Ultimately, then, what we were all trusting was other contributors' response-ability to one another and the project. To return to the theme introduced at the beginning of the book, this meant that people 'show up' for one another and 'stay with the trouble' (Haraway, 2016). For all those involved in this book this meant shifting somewhat beyond contemporary modes of authorship in neoliberal academic publishing, which largely tend to prioritize responsibility to self and one's subjects, and instead to engage our 'ability to respond, to respond to the world beyond oneself, as well as a willingness to recognize its existence' (Kuokkanen, 2010: 39). That is, a collective writing project such as this involves a response-ability for all to listen and respond to others within the group, with the intention of this project of cultivating and fostering a culture of mutual support and dialogue within the field of critical data studies. While CDS may rightly question the call to trust those that leverage data power on behalf of capitalism, we also need to foster trust and response-ability among those that in their various ways aim to counter it. Across the different chapter teams we saw traces of this response-ability beginning to take root, and our hope is that some seeds have been sown for future collaborations.

In the spirit of response-ability, here we share two responses to our reflection from authors involved in the project:

Pat Brodie (environmental data power): While I surprisingly found myself in a leading role in crafting our chapter, the guidance of Jo and the collegiality and openness of my co-authors, in spite of varying geographies, commitments, and career orientations, made the experience a nourishing one. I think our collaboration, across varied approaches but aligned by a commitment to environmental data *justice*, has together made a valuable argument about the role of critical research in crafting alternative pathways for understanding and employing environmental data power.

Karen Louise Smith with input from the chapter author team: The concept of response-ability related to the collaborative writing practices that underpin this book truly resonates with me. For much of my academic life, I have doodled in the margins of my notebook as I listen to lectures and presentations. For this book, my doodle from an initial meeting became a sketchnote which via ongoing dialogue with my co-authors turned into an annotated figure. Our braid image conveys the intent of our collaboration, built on concepts like data assemblages (Kitchin, 2014; Selwyn, 2021) and data futures (Dourish and Gómez Cruz, 2018; Ruppert, 2018).

The story behind the braid image thus demonstrates how we worked together to respond to the existing literature in critical data studies, the scholarship of the Data Power community, and finally the empirical and theoretical ideas of our co-authors through a collaborative sensemaking and writing process.

To conclude, Jo and Juliane are immensely grateful to Bristol University Press, the authors and discussants of this book to put their trust in us, and experiment with a way of doing academic response-ability and dialogue otherwise.

References

Dourish, P. and Gómez Cruz, E. (2018) Datafication and data fiction: narrating data and narrating with data. *Big Data & Society*, 5(2): 2053951718784083.

Haraway, D. (2016) *Staying with the Trouble: Making Kin in the Chthulucene*. Durham, NC: Duke University Press.

Jandrić, P., Luke, T. W., Sturm, S., McLaren, P., Jackson, L., MacKenzie, A., et al (2023) 'Collective Writing: The Continuous Struggle for Meaning-making'. In: P. Jandrić, A. MacKenzie, and J. Knox (eds) *Postdigital Research: Genealogies, Challenges, and Future Perspectives* (pp 249–293). Cham: Springer Nature Switzerland.

Kitchin, R. (2014) *The Data Revolution: Big Data, Open Data, Data Infrastructures and their Consequences*. London: SAGE.

Kuokkanen, R. (2010) The responsibility of the academy: a call for doing homework. *Journal of Curriculum Theorizing*, 26(3): 61–74.

Peters, M. A., Jandrić, P., Irwin, R., Locke, K., Devine, N., Heraud, R., et al (2016) Towards a philosophy of academic publishing. *Educational Philosophy and Theory*, 48(14): 1401–1425. https://doi.org/10.1080/00131 857.2016.1240987

Ruppert, E. (2018) 'Sociotechnical Imaginaries of Different Data Futures: An experiment in citizen data'. 3e Van Doornlezing, 14 June. www.eur.nl/ sites/corporate/files/2018-06/3e%20van%20doornlezing%20evelyn%20 ruppert.pdf

Selwyn, N. (2021) Key concepts: The 'data assemblage' [Website]. The Data Smart School. Available at: https://data-smart-schools.net/2021/05/21/ key-concept-the-data-assemblage/

Index

Page numbers in *italic* type refer to figures.